MICRO-CHP POWER GENERATION FOR RESIDENTIAL AND SMALL COMMERCIAL BUILDINGS

MICRO-CHP POWER GENERATION FOR RESIDENTIAL AND SMALL COMMERCIAL BUILDINGS

LOUAY M. CHAMRA AND PEDRO J. MAGO

Nova Science Publishers, Inc.
New York

Copyright © 2009 by Nova Science Publishers, Inc.

All rights reserved. No part of this book may be reproduced, stored in a retrieval system or transmitted in any form or by any means: electronic, electrostatic, magnetic, tape, mechanical photocopying, recording or otherwise without the written permission of the Publisher.

For permission to use material from this book please contact us:
Telephone 631-231-7269; Fax 631-231-8175
Web Site: http://www.novapublishers.com

NOTICE TO THE READER

The Publisher has taken reasonable care in the preparation of this book, but makes no expressed or implied warranty of any kind and assumes no responsibility for any errors or omissions. No liability is assumed for incidental or consequential damages in connection with or arising out of information contained in this book. The Publisher shall not be liable for any special, consequential, or exemplary damages resulting, in whole or in part, from the readers' use of, or reliance upon, this material.

Independent verification should be sought for any data, advice or recommendations contained in this book. In addition, no responsibility is assumed by the publisher for any injury and/or damage to persons or property arising from any methods, products, instructions, ideas or otherwise contained in this publication.

This publication is designed to provide accurate and authoritative information with regard to the subject matter covered herein. It is sold with the clear understanding that the Publisher is not engaged in rendering legal or any other professional services. If legal or any other expert assistance is required, the services of a competent person should be sought. FROM A DECLARATION OF PARTICIPANTS JOINTLY ADOPTED BY A COMMITTEE OF THE AMERICAN BAR ASSOCIATION AND A COMMITTEE OF PUBLISHERS.

LIBRARY OF CONGRESS CATALOGING-IN-PUBLICATION DATA

ISBN: 978-1-60456-867-7

Published by Nova Science Publishers, Inc. New York

Contents

Preface		vii
Chapter 1	Introduction	1
Chapter 2	The micro-CHP System	7
Chapter 3	Prime Movers	9
Chapter 4	Thermally Activated Devices	49
Chapter 5	Conclusions	67
References		71
Index		75

PREFACE

The traditional structure of the electrical utility market, green engineering issues, and environmental acceptability have resulted in a relatively small number of electric utilities. However, today's technology permits development of smaller, less expensive power systems, bringing in new, independent producers. Competitions from these independent producers along with the re-thinking of existing regulations have affected the conventional structure of electric utilities. The restructuring of the electric utility industry and the development of new "onsite and near-site" power generation technologies have opened up new possibilities for buildings, building complexes, and communities to generate and sell power. Competitive forces have created new challenges as well as opportunities for companies that can anticipate technological needs and emerging market trends.

Micro-cooling, heating, and power (micro-CHP) is decentralized electricity generation coupled with thermally activated components for residential and small commercial applications. A micro-CHP system consists of a prime mover, such as a reciprocating engine, which drives a generator, which produces electrical power. The waste heat from the prime mover is recovered and used to drive thermally activated components and to produce hot water or warm air through the use of heat exchangers.

Micro-CHP holds some of the answers to the efficiency, pollution, and deregulation issues that the utility industry currently faces. A review of micro-CHP systems, specific types of distributed power generation, and thermally-activated technologies are introduced and discussed in this chapter.

Chapter 1

INTRODUCTION

Micro-cooling, heating, and power (micro-CHP) is decentralized electricity generation coupled with thermally activated components for residential and small commercial applications. micro-CHP systems can simultaneously produce heat, cooling effects, and electrical power. The "micro" regime is typically designated as less than fifty kilowatts electric (< 50 kWe).

The concept of micro-CHP is illustrated in Figure 1. A prime mover, such as a reciprocating engine, drives a generator which produces electrical power. The waste heat from the prime mover is recovered and used to drive thermally activated components, such as an absorption chiller or desiccant dehumidifier, and to produce hot water or warm air through the use of heat exchangers.

Cooling, heating, and power (CHP) has proven beneficial in many industrial situations by increasing the overall thermal efficiency, reducing the total power requirement, and providing higher quality, more reliable power. Applying CHP technology to smaller scale residential and small commercial buildings is an attractive option because of the large potential market.

The residential and small commercial sectors account for 40% of the electrical usage in the U.S. As can be seen from Figure 2, the residential and small commercial sectors make up the largest portion of the utility electricity market.

The residential energy consumption is not only the largest portion of the pie, but it is also the fastest growing segment. Between 1978 and 1997, the number of U.S households has increased by over 30%. At the same time, space heating expenditures have increased by 75%, air conditioning by 140%, and water heating by 184%. The largest increase in household energy expenditures was for home appliances, which increased by 210%.

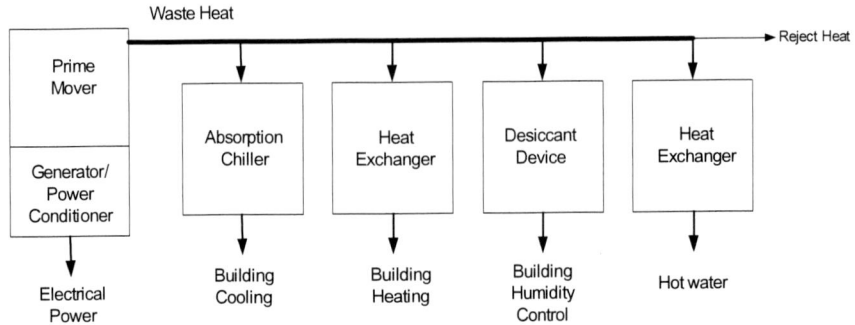

Figure 1. Schematic of a micro-CHP System.

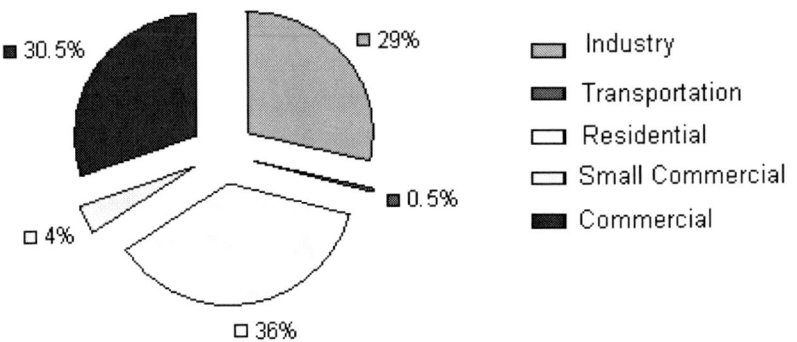

Figure 2. 2001 U.S. Electrical Consumption by Building Sector. (*Available at http://www.eia.doe.gov*).

As a result of such large increases, residential energy consumption is projected to increase by 25% from 2001-2025. The question is: Why should micro-CHP be considered a viable option to meet the needs of the U.S. residential and small commercial market?

The basis of this answer can be found by applying the "wells to wheels" analysis concept to the energy production for a single residence. The idea of "wells to wheels" is that the whole system must be considered from fuel harvesting to the energy (in some final form) that is used. In addition, each time that fuel is converted, packaged, or transported, there is an associated loss of energy. The more conversion and transportation steps in a process, the greater the associated energy losses.

Introduction 3

In the U.S. as of 2004, electricity is generated by coal (50%), nuclear (20%), natural gas (18%), hydro (7%), petroleum (3%), and various renewable energy methods (2%). The traditional method of electrical power generation and distribution is based on large, centrally-located power plants. Central means that the power plant is located on a hub surrounded by major electric load centers.

Once the electricity is produced, the power must be delivered to the end user. Delivery is achieved by a utility transmitting the electricity to a substation through a high-voltage electrical grid. At the substation, the high-voltage electricity is transformed, or stepped down, to a lower voltage to be distributed to individual customers. The electricity is then stepped down a final time by an on site transformer before being used by the customer. The number of times that the electricity must be transformed depends largely upon the distance the power is transmitted and the number of substations used in distributing the electricity.

Inefficiencies are associated with the traditional methods of electrical power generation and delivery. To begin, the majority of the energy content of the fuel is lost at the power plant through the discharge of waste heat. Traditional power plants convert about 30% of a fuel's available energy into electric power. Highly efficient combined-cycle power plants convert about 50% of the available energy into electric power. Further energy losses occur in the transmission and distribution of electric power to the individual user. Inefficiencies and pollution issues associated with conventional power plants have provided the motivation for new developments in on-site power generation. The overall efficiencies of central power generation and distributed combined-cycle power generation are shown in figures 3 and 4.

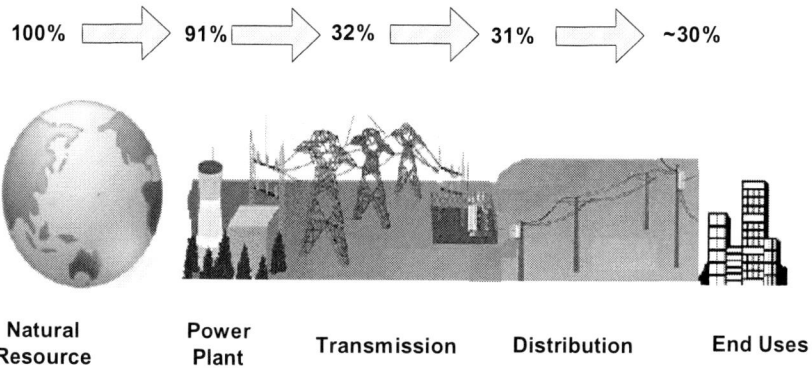

Figure 3. Efficiency of Central Power Generation.

Once the electric power reaches the end user, the electricity is used to run central heat and air conditioners, appliances, lighting, and in some cases, water heating. These are the same end uses that could be provided by a micro-CHP system at a greater overall thermal efficiency. Micro-combined heat and power units utilize waste heat while simultaneously producing electric power for a residence or building. The waste heat is used to meet space and water heating requirements and to provide cooling if an absorption chiller is incorporated into the system. Heating and cooling are major end uses of residential energy. Because the heating and cooling loads of the space are being met without total dependence on electrically-driven thermal components, the overall electric load of the residence will be reduced.

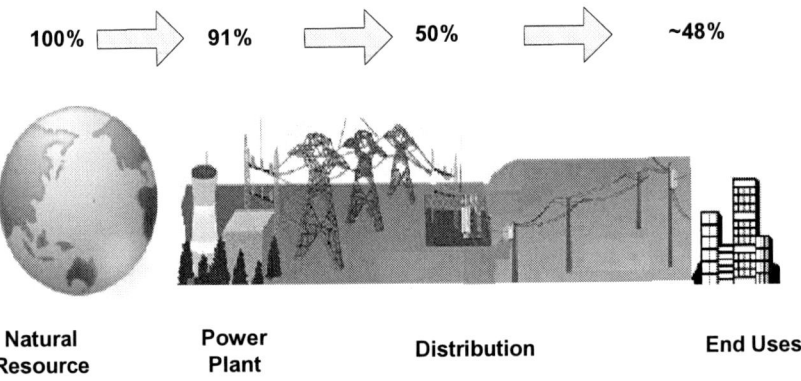

Figure 4. Efficiency of Combined-Cycle Power Generation.

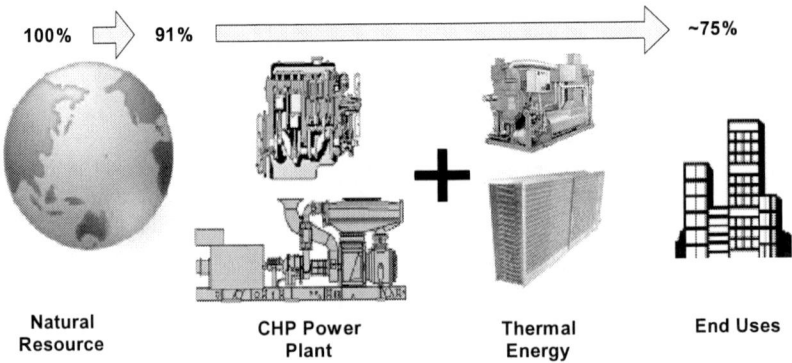

Figure 5. Efficiency of micro-CHP System.

Another advantage of micro-CHP is that there are no losses associated with power distribution and transmission as opposed to the traditional power generation method. Micro-CHP systems can utilize about 75% of the fuels available energy to provide electric and thermal energy. A micro-CHP system can produce an overall efficiency of about 75% while a modern combined-cycle power plant will have an overall efficiency of around 50%. The overall efficiencies of a micro-CHP system is shown in Figure 5.

Larger homes, higher energy costs, volatile fuel markets, electricity blackouts, power security, power quality, and increasing concern for environmental issues have all helped open the door for micro-CHP. RKS, a leading market research firm, found that more than 38% of high-income households, (i.e., incomes greater than $50,000) are interested in generating their own electricity. (Micro-CHP Technologies Roadmap, U.S. DOE, 2003).

HISTORY

Combined heat and power generation, or cogeneration, is a well established concept dating back to the 1880s when steam was a primary source of energy in industry and electricity was beginning to be used for both power and lighting. As electrical power and electrical motors became more widely used, steam driven mechanisms were replaced creating a transition from mechanically powered systems to electrically powered systems. In the early 1900s, an estimated 58% of the total power generated in the United States by on-site industrial power plants was cogenerated power.

The development of central power plants and reliable utility grids drove electricity costs down, and industrial plants began buying electricity from utility companies and ceased generating their own power. On-site industrial cogeneration declined in the United States and accounted for only 15% of total electricity generation by 1950 and dropped to about 5% by 1974 (Knight and Ugersal, 2005). Increasing regulatory policies, low fuel costs, and advances in technology also contributed to the decline of cogeneration.

In the last forty years systems that are efficient and have the ability to utilize alternative fuels have begun to appear because of energy price increases and the uncertainty of fuel supplies. In addition, CHP has gained attention because of decreased fuel consumption and lower emissions. Today, many industrialized countries are taking leading roles in establishing and promoting the use of cogeneration in the industrial, residential, and other market sectors.

Chapter 2

THE MICRO-CHP SYSTEM

Micro-cooling, heating, and power combines distributed power generation with thermally activated components to meet the cooling, heating, and power needs of residential and small commercial buildings. The success of CHP systems for large-scale application coupled with the development of power generation equipment and thermally activated components on a smaller scale have contributed to the development of micro-CHP applications. Distributed power generation technologies and thermally activated components will be introduced and briefly discussed.

DISTRIBUTED POWER GENERATION

A number of technologies are commercially available or under development for generating electric power (or mechanical shaft power) onsite or near site where the power is used. Distributed power generation is a required component of micro-CHP systems. Fuel cells, reciprocating engines, Stirling engines, Rankine cycle engines, and microturbines are prime movers that have the most potential for distributed power generation for micro-CHP systems.

When discussing various prime movers for micro-CHP systems, a primary method of comparison is to examine the efficiency of each prime mover. The efficiency of a micro-CHP system is measured as the fraction of input fuel that can be recovered as power and heat. The remaining energy is rejected as low-temperature heat. There are three primary efficiencies that are associated with micro-CHP systems: electrical efficiency, thermal efficiency, and overall efficiency. These efficiencies are defined as

$$\text{Electical efficiency} = \frac{\text{electrical output}}{\text{fuel input}} \qquad (1)$$

$$\text{Thermal efficiency} = \frac{\text{Thermal output}}{\text{fuel input}} \qquad (2)$$

$$\text{Overall efficiency} = \frac{\text{useful thermal} + \text{electrical output}}{\text{fuel input}} \qquad (3)$$

Chapter 3

PRIME MOVERS

RECIPROCATING ENGINES

Reciprocating engines can be used to produce shaft power. The shaft power can then be used to drive a generator to produce electrical power. The shaft power can also be used to operate equipment such as compressors and pumps. The application of reciprocating engines is widespread and highly developed. Reciprocating engines use natural gas, propane, gasoline, diesel and biofuels to produce 0.5 kW to 10 MW of power. A diesel fuel engine generator set is shown in Figure 6.

Reciprocating engines exhibit characteristics that are advantageous for micro-CHP applications. Reciprocating engines used for power generation have proven reliability, good load-following characteristics, low capital cost, fast startup, and significant heat recovery potential. Recent advances in combustion design and exhaust catalyst have also helped reduce overall emissions of reciprocating engines. Currently, reciprocating engines are the most widely used distributed energy technology. Typical electrical conversion efficiencies are in the range of 25% to 40%. The overall thermal efficiencies of these systems increase with the incorporation of thermally activated components.

The thermal energy in the engine cooling system and exhaust gases from reciprocating engines can often be recaptured and used for space heating, for hot water heating and for driving thermally activated components. Shaft power from the engine can also be used to power thermal components, such as gas vapor compression chillers. Such chillers are very similar to electric-driven chillers with the exception that the compressor is driven by the reciprocating engine rather than an electric motor.

Figure 6. Model D13-2, 12-kW Diesel Engine Generator Set from Caterpillar (*Available at http://www.cat.com*).

**Table 1. Overview of Reciprocating Engine Technology
(Available at http://www.energy.ca.gov/distgen/)**

Reciprocating Engines Overview	
Commercially Available	Yes
Size Range	0.5 kW – 7 MW
Fuels	Natural gas, diesel, landfill gas, digester gas
Efficiency	25 – 45%
Environmental	Emission controls required for NOx and CO
Other Features	Cogeneration (some models)
Commercial Status	Products are widely available

Emissions of reciprocating engines tend to be higher than that of other distributed generation equipment. Due to the emissions and noise emitted by these engines, care must be exercised in the location of the engine with respect to the occupants of the building. In some areas, local air quality standards may limit the use of reciprocating engines.

Application

Reciprocating engine generator sets are the most common and most technically mature of all distributed energy resources (DER) technologies. Reciprocating engines can be used for a variety of applications due to their small size, low unit costs, and useful thermal output. Applications for reciprocating engines in power generation include continuous or prime-power generation, peak shaving, back-up power, premium power, remote power, standby power, and mechanical drive use. An overview of reciprocating engine characteristics is presented in table 1.

Reciprocating engines are an ideal candidate for applications in which there is a substantial need for hot water or low pressure steam. The thermal output can be used in an absorption chiller to provide cooling. Comparatively low installation costs, suitability for intermittent operation, and high temperature exhaust make combustion engines an attractive option for micro-CHP. Internal combustion engines utilize proven technologies with a well established infrastructure for mass production and marketing. The development of combustion engines has also formed a maintenance infrastructure with certified technicians and relatively inexpensive and available parts are available. Due to the long history and widespread application, internal combustion engines are a more developed technology than most prime movers considered for micro-CHP.

Heat Recovery

Traditional large-scale electric power generation is typically about 30% efficient, while combined cycle plants are typically 48% efficient. In either case, the reject heat is lost to the atmosphere with the exhaust gases. In an internal combustion engine, heat is released from the engine through coolant, surface radiation, and exhaust. Engine-driven micro-CHP systems recover heat from the jacket water, engine oil, and engine exhaust. Low pressure steam or hot water can be produced from the recovered heat, and can be used for space heating, domestic hot water, and absorption cooling.

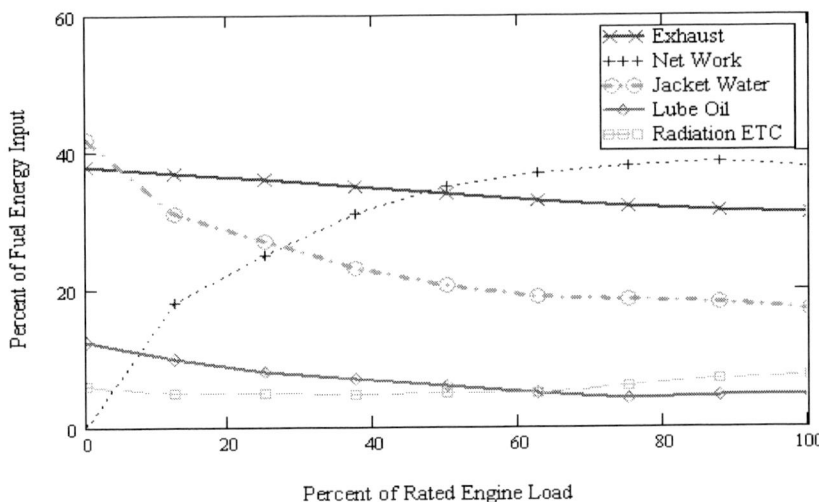

Figure 7. Heat Balance for a Representative Reciprocating Engine (Knight and Ugersal, 2005).

Heat from the engine jacket coolant is capable of producing 200 F (93 C)hot water and accounts for approximately 30 % of the energy input from the fuel. Engines operating at high pressure or equipped with ebullient cooling systems can operate at jacket temperatures of up to 265 F (129 C). Engine exhaust heat can account for 10 – 30 % of the fuel input energy and exhaust temperatures of 850 F –1200 F (455 C – 649 C) are typical. Because exhaust gas temperatures must be kept above condensation thresholds, only a portion of the exhaust heat can be recovered. Heat recovery units are typically designed for a 300 F – 350 F exhaust outlet temperature to avoid corrosive effects of condensation in the exhaust piping. Low-pressure steam (~15 psig) and 230 F (110 C) hot water are typically generated using exhaust heat from the engine. The combined heat recovery of the coolant and exhaust in conjunction with the work produced by combustion can utilize approximately 70 – 80% of the fuel energy. Figure 7 shows a heat balance for a representative reciprocating engine.

Cost

Reciprocating internal combustion (IC) engines are the traditional technology for emergency power all over the world. They have the lowest first costs among DER technologies. The capital cost of a basic gas-fueled generator set (genset)

The waste heat from a microturbine is primarily in the form of hot exhaust gases. This heat is suitable for powering a steam generator, indirect heating of a building, allocation to thermal storage devices, or use in heat-driven cooling systems. Most designs incorporate recuperators that limit the amount of heat available for micro-CHP applications.

Microturbines have relatively low emissions and noise and also have low maintenance costs. Another advantage is that microturbines are relatively small in size or footprint. The fuel flexibility and quantity of hot exhaust gases make microturbines an advantageous technology for micro-CHP and cogeneration applications. The Capstone C30, a 30-kW microturbine, is pictured in Figure 8.

Application

Markets for microturbines include commercial and light industrial facilities. Microturbines can be used for stand-by power, power quality and reliability, peak shaving, and cogeneration applications. In addition, because microturbines are being developed to utilize a variety of fuels, microturbines are used for resource recovery and landfill gas applications. Microturbines are well-suited for small commercial establishments such as restaurants, hotels, motels, small offices, retail stores.

The development of the microturbine technology for transportation applications is also in progress. Automotive companies are interested in microturbines to provide a lightweight and efficient fossil-fuel-based energy source for hybrid electric vehicles, especially buses. Microturbines are also being developed to utilize a variety of fuels and are being used for resource recovery and landfill gas applications.

Heat Recovery

The waste heat from a microturbine is primarily in the form of hot exhaust gases. This heat is suitable for powering a steam generator, heating of a building, allocation to thermal storage devices, or use in absorption cooling system. However, most designs incorporate a recuperator that limits the amount of heat available for micro-CHP applications.

The manner in which the waste heat can be used depends upon the configuration of the turbine system. In a non-recuperated turbine, the exhaust gas typically exits at a temperature between 1000 – 1100 F (538 – 594 C). A recuperated turbine can provide waste heat for heating and operating an absorption cooling system at exhaust temperatures around 520 F (271 C). The recovered heat can also be used to drive a desiccant dehumidification device. The use of the recovered heat influences the selection of the microturbine with or without a recuperator.

Cost

The capital costs of microturbines range from $700 - $1,100/kW when mass produced. These costs include all hardware, associated manuals, software, and initial training. Adding heat recovery components increases the cost by $75-$350/kW. Installation and site preparation can increase the capital costs by 30-50%. Manufacturers are striving for future capital costs of microturbines to be below $650/kW. This goal appears feasible if the market expands and sales volumes increase.

With fewer moving parts, vendors hope their microturbines can provide higher reliability and require less maintenance than conventional reciprocating engine generators. The single-shaft design with air bearings will not require lubricating oil or water. Microturbines that use lubricating oil should not require frequent oil changes as the oil is isolated from the combustion products.

Manufacturers expect microturbines to require maintenance once-a-year when the technology matures and are targeting maintenance intervals of 5,000 – 8,000 hours. Actual maintenance costs and intervals for mature microturbines are less well known since there is a limited base of empirical data from which to draw conclusions. Forecasted maintenance costs for microturbines range from $0.005 – $0.016 per kWh, slightly lower than the costs for small reciprocating engine systems.

Advantages and Disadvantages

The operation of a microturbine offers several advantages. Microturbines have fewer moving parts than IC engines. The limited number of moving parts and the low lubrication requirements allow microturbines long maintenance intervals. Accordingly, microturbines have lower operating costs in terms of cost per kilowatt of power produced. Another advantage of microturbines is their relatively small size for the amount of power that is produced. Microturbines are also light weight and have relatively low emissions. Potentially, one of the greatest advantages of microturbines is their ability to utilize a number of fuels, including waste fuels or biofuels. Microturbines have great potential in cogeneration applications because microturbines produce a large quantity of clean, hot exhaust gases compared to other distributed generators.

The primary disadvantages of microturbines are that they have a low fuel to electrical efficiency. Also, with higher elevation and increased ambient temperatures, microturbines experience a loss of power output and efficiency. The ambient temperature directly affects the temperature of the air at the intake. A gas turbine will operate more effectively when colder air is available at the intake. A gas turbine cycle must compress the inlet air and the greater the compression, the greater efficiency. Another potential disadvantage is that microturbines experience more efficient operation and require less maintenance when operated continuously.

STIRLING ENGINES

The Stirling engine is a type of external combustion piston engine which uses a temperature difference to produce motion. The cycle is based on the behavior of a fixed volume of gas. The heat source used to provide the temperature difference can be supplied by a wide variety of fuels or solar energy. The Stirling engine has only seen use in specific and somewhat limited applications. However, recently many companies have begun research and development related to Stirling engines due to their potential for micro-CHP applications and solar power stations.

Stirling engines typically have an electrical efficiency in the range of 12% to 25%. This efficiency can be increased with the use of recuperators. The operation of a Stirling engine requires that one side of the engine remain hot while the other side remains cool. This requirement makes heat recovery an integral part of the operation of a Stirling engine. Heat can be recovered from dissipation of the heat source and through the use of heat exchangers on the cool side of the engine. Stirling engines have low emissions and create low noise levels. These engines are also mechanically simple, and because there is no internal combustion, the maintenance requirements of Stirling engines are relatively low. However, due to design, Stirling engines are heavy and large for the amount of power generated. Stirling engines also have one of the higher capital costs of distributed power generation technologies. The SOLO 9-kW Stirling engine based micro-CHP unit is shown in Figure 9.

Stirling engines use a displacer piston to move the enclosed gas back and forth between the hot and cold reservoirs. The gas expands at the hot reservoir and displaces a power piston, producing work while at the same time forcing the gas to move to the cold reservoir. At the cold reservoir, the gas contracts, retrieving the power piston and closing the cycle. The operation of a Stirling engine can be best understood by examining the operation of a two-cylinder (or alpha) Stirling engine. A step-by-step diagram of a two-cylinder Stirling engine and further explanation on the operation of the Stirling engine are shown in Figure 10. In the two-cylinder Stirling engine, one cylinder is kept hot while the other is kept cool. In Figure 10, the lower-left cylinder is heated by burning fuel. The other cylinder is kept cool by an air cooled sink. In a two-cylinder Stirling engine each piston acts as both a power piston and displacer piston.

Figure 9. SOLO 9-kW Stirling Engine (Available at http://www.stirling-engine.de/engl/index.html.

Application

Stirling engines are an old technology; however their use became limited with the improvement of steam engines and the invention of the Otto cycle engine. Recent interest in distributed energy has revived interest in Stirling engines. Stirling engines can be used in a variety of applications due to their thermal output, simple operation, low production costs, and relatively small size. Applications for Stirling engines include power generation units for space crafts and vehicles, small aircraft, refrigeration, micro-CHP, solar dish application, and small scale residential or portable power generation. An overview of Stirling engine characteristics is presented in table 2.

Commercially available Stirling engines can produce between 1 kW – 25 kW. Stirling engine technology has been widely used in England and Europe with great success, particularly in the micro-CHP arena. However, a very small

Prime Movers 19

percentage of the world's electrical capacity is currently provided through the use of Stirling engines.

Table 2. Overview for Stirling Engine Technology
(Available at http://www.energy.ca.gov/distgen)

Stirling Engine Overview	
Commercially Available	Yes, mostly in Europe
Size Range	<1 kW - 25 kW
Fuel	Natural gas primarily but broad fuel flexibility is possible
Efficiency	12 – 20% (Target: >30%)
Environmental	Potential for very low emissions
Other Features	Cogen (some models)

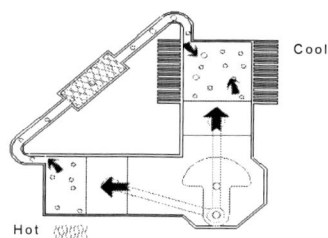

Step 1: Compression. At this point, the majority of the gas has been shifted to the cool cylinder. As the gas cools and contracts, both pistons are drawn outward.

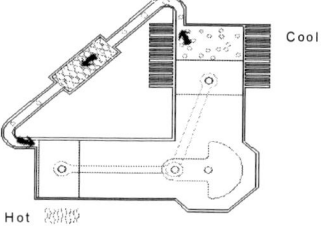

Step 2: Heat Transfer. The now contracted gas is still located in the cool cylinder. Flywheel momentum carries the crank another 90 degrees, transferring the gas to back to the hot cylinder.

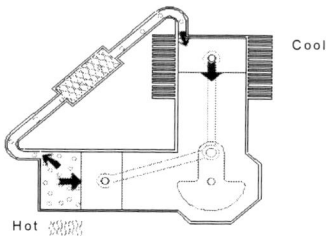

Step 3: Expansion. Now, most of the gas in the system has just been driven into the hot cylinder. The gas heats and expands driving both pistons inward.

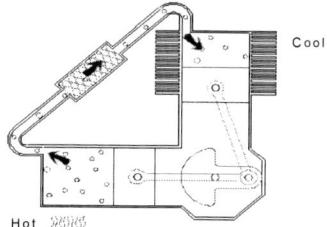

Step 4: Heat Transfer. At this point, the gas has expanded (about 3 times in this example). Most of the gas (about 2/3rds) is still located in the hot cylinder. Flywheel momentum carries the crankshaft the next 90 degrees, transferring the bulk of the gas to the cool cylinder to complete the cycle.

Figure 10. Two-cylinder Stirling Engine Diagram (*Available at www.keveny.com/ vstirling.html*).

Heat Recovery

Traditional large-scale electric power generation is typically about 30% efficient, while combined cycle power plants are approximately 48% efficient. In either case, the reject heat is lost to the atmosphere with the exhaust gases. Stirling engines are typically in the range of 15 – 30% efficient, with many reporting efficiencies of 25 – 30%. The overall efficiencies of these systems can be greatly increased by recovering the waste heat.

A high percentage of the Stirling engine heat losses will go to the cooling fluid instead of into the exhausts, which makes the Stirling engine suitable for combined heat and power generation. Typical operating temperatures range from 1200 – 1470 F (650 – 800 C), resulting in electrical engine conversion efficiencies of around 30% to 40% when a recuperator is included in the engine system. These high operating temperatures can convert into high quality waste heat. The reject heat can be recaptured through piping the cooling fluid through a heat exchanger and by ducting the exhaust gases through a heat exchanger to produce hot water.

Cost

The capital costs of Stirling engines are comparably high and vary greatly depending upon manufacturer ($2,000 - $50,000). And, overall cost increases with size. However, several companies have targeted Stirling technology for micro-CHP units and have achieved relative success. These companies include PowerGen, WhisperTech, Sunpower, and ENATEC.

Advantages and Disadvantages

Some advantages of Stirling engines are:

- The burning of the fuel air mixture can be more accurately controlled due to the external heat source.
- Emission of unburned fuel can be eliminated as a continuous combustion process can be used to supply heat.
- Less lubrication is required leading to greater periods between overhauls because the bearings and seals are placed on the cool side.
- Simplicity of design; no valves are needed, fuel and intake systems are very simple.

- Low noise and vibration free operation.
- Low maintenance and high reliability.
- Multi-fuel capability.
- Long service life.

Some disadvantages of Stirling engines are:

- High costs.
- Low efficiencies.
- Require both input and output heat exchangers which must withstand the working fluid pressure and resist corrosion effects.
- Relatively large for the amount of power they produce due to the heat exchangers.
- Cannot experience instantaneous start-up.
- Power output is relatively constant and rapid change to another level is difficult to achieve.

Rankine Cycle Engines

Rankine cycle engines are based upon the well known thermodynamic cycle that is used in most commercial electric power plants. The shaft power from a Rankine cycle engine is used to drive an electric generator in the same manner as reciprocating or Stirling engines. Rankine cycle engines have relatively low electrical conversion efficiency. However, as micro-CHP technologies are designed to follow the thermal load, this low electrical efficiency becomes less of a drawback because significant thermal energy that can be recovered from a Rankine cycle engine. The durability and performance characteristics of Rankine cycle engines are also well known, and low production costs are a potential benefit.

The construction of a Rankine cycle engine allows heat to be recaptured easily through the use of a condenser, which is already a component in the engine cycle. Currently, Rankine cycle engines for micro-CHP applications are in the development stage. As a result, cost and specific performance characteristics are not yet defined. A Cogen Microsystems 2.5-kW micro-CHP unit based on a Rankine cycle engine is pictured in Figure 11.

Application

The advantage of the Rankine cycle for power plants is that the working fluid is a liquid. Many times this liquid is water, which is a cheap and readily available resource. Currently, companies such as the Baxi Group, Enginion, and Cogen Microsystems are exploring the possibility of using Rankine Cycle engines for micro-CHP.

Heat Recovery

As the Rankine cycle is a closed-loop, which incorporates a condenser, heat recovery can be achieved easily at the condenser. However, as most of the Rankine cycle engine technologies for micro-CHP are still under development, the quality and quantity of heat that can be recaptured is currently not well defined.

Figure 11. Cogen Microsystems 2.5-kW Rankine Cycle micro-CHP Unit (*Available at www.cogenmicro.com*).

Cost

Unfortunately, little is known about potential costs of small-scale Rankine cycle units for micro-CHP. As more of these technologies complete the field trial stage of development, more information concerning capital and maintenance costs will be available.

FUEL CELLS

Fuel cells are electrochemical energy conversion devices that produce electrical power rather than shaft power. Unlike the technologies discussed previously, fuel cells have no moving parts and, thus, no mechanical inefficiencies. Four major types of fuel cells will be discussed: proton exchange membrane (PEMFC), solid oxide (SOFC), phosphoric acid (PAFC), and molten carbonate (MCFC) fuel cells. Each of these fuel cell types operate differently and exhibit different performance characteristics.

In general terms, fuels cells combine a hydrogen based fuel input and gaseous stream containing oxygen in the presence of a catalyst to initiate a chemical reaction. The products of this reaction vary for each type of fuel cell but typically are electrical power, heat, and water. In some instances, other product gases such as carbon dioxide are formed. As a pure hydrogen-rich fuel is required by most fuel cells, hydrogen reformers are often included in a fuel cell system.

Like batteries, fuel cells produce direct current (DC) electrical power. This requires that an inverter and power conditioner be used to transform the DC current into alternating current (AC) at the appropriate frequency for use in the majority of applications.

Fuel cells can achieve high electrical efficiencies as compared to other distributed power generation equipment. Fuel cells exhibit quiet operation and low emissions. Also, the absence of mechanical components decreases maintenance. Unfortunately, the costs of fuel cells are relatively high as compared to other technologies. The fuel flexibility of fuel cells is also low as very pure streams of hydrogen are the only suitable fuel for certain types of fuel cells. In some instances, the energy required to reform the input fuel greatly decreases the overall efficiency of a fuel cell system. Still, fuel cells are a promising technology that hold potential for micro-CHP applications. A Plug Power fuel cell is shown in Figure 12.

Figure 12. Plug Power Fuel Cell Unit (*Available at www.plugpower.com*).

Typically, a fuel cell electrochemically reacts hydrogen and oxygen to form water and in the process produces electricity. The process consists of a hydrogen-based input fuel passing over an anode where a catalytic reaction occurs, splitting the fuel into ions and electrons. Consider the reactions at the anode and cathode of an acid electrolyte fuel cell. At the anode, the hydrogen gas ionizes, releasing electrons and creating H^+ ions in an exothermic reaction. The reaction is

$$2H_2 \rightarrow 4H^+ + 4e^-$$

Ions pass from the anode, through the electrolyte, to an oxygen-rich cathode. At the cathode, oxygen reacts with the electrons taken from the electrode and the H^+ ions that have traveled through the electrolyte to form water according to the following reaction:

$$O_2 + 4H^+ + 4e^- \rightarrow 2H_2O$$

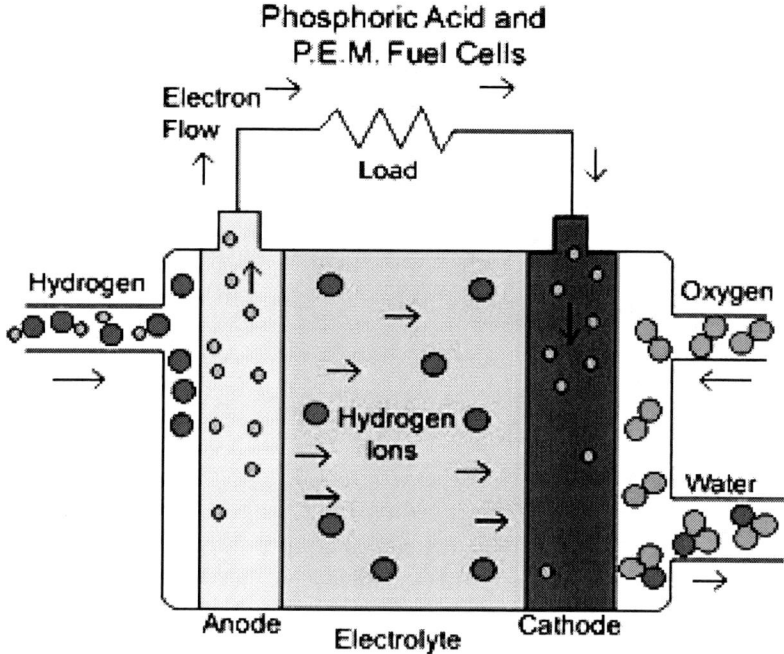

Figure 13. PAFC Electrochemistry (Available at http://www.fctec.com/fctec).

For the reactions to occur simultaneously, electrons produced at the anode must pass through an electric circuit to the cathode while the H⁺ ions pass through the electrolyte. A schematic diagram of the reaction occurring in a phosphoric acid fuel cell (PAFC) is illustrated in Figure 13.

The performance of a fuel cell can be analyzed by using the Gibbs function or Gibbs free energy of reaction. The Gibbs free energy of reaction is an indication of the maximum work that can be achieved from combining two substances in a chemical reaction.

The maximum theoretical work (W_{MAX}) is the difference in the Gibbs free energy (G) of the reactants and the products of the chemical reaction.

$$W_{MAX} = G_{react} - G_{prod} \tag{4}$$

The Gibbs free energy is a function of the enthalpy (h), entropy (s), and temperature (T) associated with the chemical compound of interest at a specific state of matter and is defined as

$$G = h - T \cdot s \tag{5}$$

The temperature-entropy product represents the loss due to changes in entropy. The maximum work that can be achieved by a fuel cell is

$$W_{MAX} = \left(h_{react} - h_{prod} \right) - T \cdot \left(s_{react} - s_{prod} \right) \tag{6}$$

and the thermal efficiency (η_{th}) is defined as

$$\eta_{th} = \frac{G_{react} - G_{prod}}{h_{react} - h_{prod}} \tag{7}$$

The thermal efficiency of a fuel cell typically ranges from 82 – 94% in ideal cases. The thermal efficiency of a fuel cell as defined in Equation 4-45 is not the actual electrical conversion efficiency achieved by the fuel cell. Due to ohmic losses resulting from concentration polarization and activation polarization the electrical conversion efficiency of a fuel cell (η_{fc}) generally falls between 40 – 60%. The ohmic losses are accounted for using a voltage efficiency (η_v) and a current efficiency (η_i). The electrical conversion efficiency of a fuel cell is defined as the product of the thermal, voltage, and current efficiencies.

$$\eta_{fc} = \eta_{th} \cdot \eta_v \cdot \eta_i \tag{8}$$

The components needed for the operation of a fuel cell system vary depending on the type of fuel cell used and the fuel. Common major components for a fuel cell system include a fuel reformer (processor), the fuel cell "stack," and a power conditioner. The fuel reformer, also known as a fuel processor, generates hydrogen-rich gas from the supply fuel and removes poisons from the fuel stream. Poisoning and fuel flexibility are two major considerations when selecting a fuel cell. Nitrogen, phosphorous, arsenic, antimony, oxygen (in specific instances), sulfur, selenium and tellurium (Group VA and VI A on the Periodic table) are typical poisons for fuel cells. Table 3 displays the fuel requirements of each fuel cell type.

Fuel reformers are needed because the fuel is often in the form of hydrocarbons, such as methane, and the hydrogen concentration is not at a level suitable for operation. The fuel cell "stack" consists of layers of the cathode, anode, and electrolyte mentioned earlier. A power conditioner converts the direct current (DC) electricity generated by the fuel cell into alternating current (AC) electricity at the appropriate voltage and frequency.

Fuel cells are categorized by the electrolyte used to transport ions between the cathode and the anode (or vise versa). Fuel cell technologies include molten carbonate (MCFC), alkaline (AFC), proton exchange membrane (PEMFC), solid oxide (SOFC), direct methanol (DMFC), and phosphoric acid (PAFC). While PEMFC are arguably the most advanced of the fuel cell technologies, the low operating temperatures are not ideal for CHP applications, particularly if the waste heat is required to drive thermally activated components. Because high temperature or high quality waste is heat needed for micro-CHP applications, the solid oxide (SOFC) and the molten carbonate (MCFC) configurations are two likely candidates. Direct methanol fuel cells also hold some potential because a hydrogen reformer is not required, thereby reducing the system costs. Currently, the PEMFC and the SOFC are regarded as having the most potential for micro-CHP. Table 4 shows the characteristics of the five major types of fuel cell technologies.

PEMFC

Proton exchange membrane fuel cells (PEMFC) use an ion-conducting polymer as the electrolyte. The electrolyte works well at low temperatures, typically around 175 – 212 F (80 – 100 C), which allows for fast start-up. The polymer membrane construction varies depending on manufacturer; however a standard practice uses a modified polymer known as polytetrafluoroehtylene (PTFE), or as it is commonly know, Teflon®. Because the electrolyte is a solid polymer, electrolyte loss is not an issue with regard to stack life. Most PEMFC's use platinum as the catalyst for both the anode and the cathode. Hydrogen is used as the fuel at the anode, and air or oxygen is supplied to the cathode. The reactions that occur within a generic PEMFC are as follows:

Table 3. Fuel Requirements for Fuel Cells (Laramie et al., 2003)

Gas Species	PEMFC	PAFC	MCFC	SOFC
H_2	Fuel	Fuel	Fuel	Fuel
CO	Poison (>10 ppm)	Poison (>0.5%)	Fuel	Fuel
CH_4	Diluent	Diluent	Diluent	Diluent
CO_2 and H_2O	Diluent	Diluent	Diluent	Diluent
S (as H_2S and COS)	Few studies, to date	Poison (>50 ppm)	Poison (>0.5 ppm)	Poison (>1.0 ppm)

Table 4. Characteristics of Fuel Cells (Available at http://www.oit.doe.gov)

	Alkaline (AFC)	Proton Exchange Membrane (PEM)	Phosphoric Acid (PAFC)	Molten Carbonate (MCFC)	Solid Oxide (SOFC)
Electrolyte	Alkaline lye	Perfluorated sulphonated polymer	Stabilized phosphoric acid	Molten carbonate solution	Ceramic solid electrolyte
Typical Unit Sizes (kW)	<<100	0.1-500	5-200 (plants up to 5,000)	800-2000 (plants up to 100,000)	2.5-100,000
Electric Efficiency	Up to 70%	Up to 50%	40-45%	50-57%	45-50%
Installed Cost ($/kW)		4,000	3,000-3,500	800-2,000	1,300-2,000
Commercial Availability	Not for CHP	R&D	Yes	R&D	R&D
Power Density lbs/kW ft^3/kW		8-10 ~0.2	~25 0.4	~60 ~1	~40 ~1
Heat Rejection (Btu/kWhr)		1640 @ 0.8 V	1880 @0.74V	850 @0.8V	1780 @0.6V
Electric/ Thermal Energy		~1	~1	Up to 1.5	Up to 1.5
Oxidation Media	Oxygen	Oxygen from Air	Oxygen from Air	Oxygen from Air	Oxygen from Air
Cooling Medium		Water	Boiling Water	Excess Air	Excess Air
Fuel	H_2	H_2 and reformed H_2	H_2 reformed from natural gas	H_2 and CO reformed from natural gas or coal gas	H_2 and CO reformed from natural gas or coal gas
Operating Temp (F)	160-210	120-210	320-410	1250	1500-1800
Operating Pressure (psig)		14.7-74	14.7-118	14.7-44	14.7->150
Applications	Space and military (today)	Stationary power (1997-2000) Bus, railroad, automotive propulsion (2000-2010)	Stationary power (1998) Railroad propulsion (1999)	Stationary power (2000->2005)	Stationary power and railroad propulsion (1998->2005)

Anode Reaction: $H_2 \Leftrightarrow 2H^+ + 2e^-$

Cathode Reaction: $O_2 + 4H^+ + 4e^- \Leftrightarrow 2H_2O$

Overall Reaction: $2H_2 + O_2 \Leftrightarrow 2H_2O$

The electrode reactions are shown in Figure 14. Hydrogen ions and electrons are produced from the fuel gas at the anode. The hydrogen ions travel through the electrolyte to the cathode. Electrons pass through an outside circuit to join the hydrogen ions and oxygen atoms at the cathode to produce water and product gases. The solid electrolyte does not absorb the water. The operation of a PEMFC requires a certain level of water be maintained in the stack; however, too much water, or a "flood" of water, will shut the fuel cell down. The optimum water level creates water management issues in that the amount of water remaining in the fuel cell stacks and the amount of water rejected must be carefully controlled.

The primary advantage of the PEMFC is that extensive development has resulted in increased electrical efficiency and decreased size and material requirements. However, there are several disadvantages of using a PEMFC for micro-CHP applications. The operating temperature is too low to drive the thermally activated components of the micro-CHP system. Water management and humidity control are critical issues in the operation of a PEMFC. There must be sufficient water content in the polymer electrolyte as the proton conductivity is directly proportional to the water content. However, there must not be so much water that the electrodes are bonded to the electrolyte, blocking the pores in the electrodes or the gas diffusion layer. Also, due to poisoning issues, relatively pure hydrogen is needed at the anode and a hydrogen reformer is necessary, which increases the overall costs of the system.

The American Council for an Energy Efficient Economy (ACEEE) Emerging Technologies & Practices (2004) performed a theoretical study on a residential CHP system using a 2-kW PEMFC as the power generation device. The economics indicated that with the estimated costs of the PEMFC unit and maintenance cost, the cost of the electricity generated would be $0.18/kWh. Therefore the installation of the 2-kW PEMFC unit would only be advantageous in an area that has an electrical cost higher than $0.18/kWh, or in an area where grid electricity is unavailable.

Thermally activated components were included only implicitly in the study, and the systems were not sized large enough to have excess power either to store or sell to the electricity grid. The study listed several major market barriers: dwindling natural gas supplies, introducing and integrating a new technology and overcoming the inertia of the established market, and uncertain system reliability. Also to be considered is how and who will provide system maintenance.

The Entergy Centre of the Netherlands (ECN) experimented with a 2 kW PEMFC micro-CHP system. The system uses natural gas, and the fuel cell operates at approximately 65 C. For system operation, natural gas is desulphurised and converted to hydrogen rich gas in the reformer. The study showed that a start-up time of 2.5 hours was needed when starting from cold conditions to steady operating conditions where the rejected heat could be used to aid the reforming process. This start-up time was reduced to 45 minutes when transitioning from hot stand-by conditions. Characteristics demonstrated by the system include:

- Gross (electrical + thermal) efficiency of fuel processor varies from 70% at 1 kW to 78% at full load (10 kW)
- Stack electrical conversion efficiency of 40% at 2 kW and 42% at 1 kW
- Recovered 53% of the waste heat (LHV basis)

The results of this study predicted a payback period of five years for a 1 kW micro-CHP system costing 1000 – 1500 EUR ($1300 - $2000), taking into account the market value of Dutch natural gas, electric rates in the Netherlands, and local energy tariffs. *Note that system costs of $1300-$2000 is a theoretical value the manufacturers hope to achieve.*

SOFC

Solid oxide fuel cells use an oxide ion-conducting ceramic material as the electrolyte. The anode of a SOFC is usually a cermet composed of nickel and yttria-stabilised zirconia. A cermet is a mixture of ceramic and metal. The cathode is a porous structure typically made of lanthanum manganite. All of the materials used to construct a SOFC are solid state. SOFC's operate in the temperature range of 800 – 1100 C. Either hydrogen or methane can be supplied at the anode, and a SOFC can accommodate both oxygen and air at the cathode. The reactions that occur within a generic SOFC are illustrated in Figure 14.

The reactions that occur within a generic SOFC are as follows:

Anode Reaction: $H_2 + O^{2-} \Leftrightarrow H_2O + 2e^-$
$CO + O^{2-} \Leftrightarrow CO_2 + 2e^-$
$CH_4 + 4O^{2-} \Leftrightarrow 2H_2O + CO_2 + 8e^-$
Cathode Reaction: $O_2 + 4e^- \Leftrightarrow 2O^{2-}$
Overall Reaction: $2H_2 + O_2 \Leftrightarrow 2H_2O$

In the SOFC reactions, hydrogen or carbon monoxide in the fuel stream reacts with oxide ions traveling through the electrolyte. These reactions produce water and carbon dioxide and disperse electrons to the anode. The electrons pass through the exterior load and return to the cathode. At the cathode, the electrons are used to ionize the oxygen molecules from the air. The oxide ions then enter the electrolyte and the process begins again.

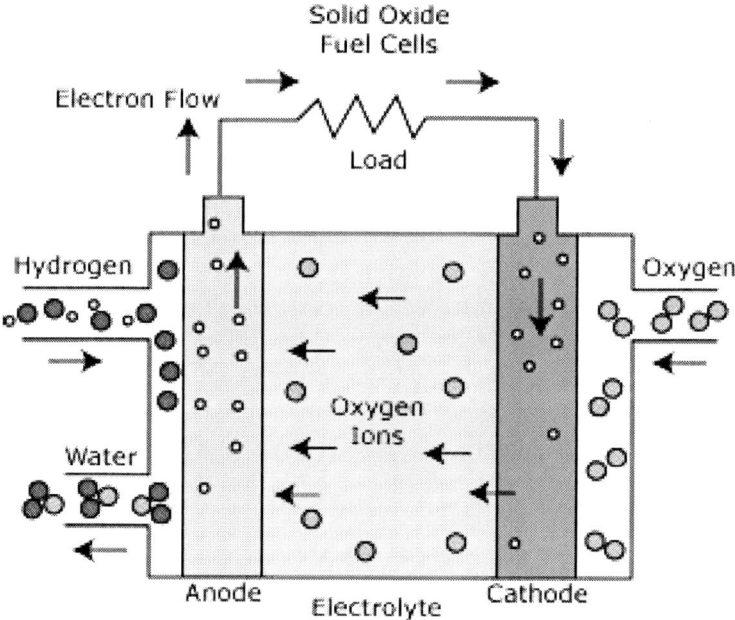

Figure 14. SOFC Reactions (Available at http://www.fctec.com/fctec).

Solid oxide fuel cells have several advantages from an micro-CHP point of view. First, the high operating temperatures are attractive for driving thermally activated components and for heating domestic hot water. The construction negates any electrolyte management issues or water management difficulties. The presence of nickel at the cathode can be used as an internal reforming catalyst, eliminating the need for a reformer and reducing system cost.

The SOFC has challenges and disadvantages. The high temperatures result in construction and material difficulties. The elevated temperatures also decrease the open-circuit voltage achieved by a SOFC because the Gibbs free energy of formation of the products (water) tends to be less negative. The reduced open-circuit voltage leads to a decrease in electrical efficiency. However, the advantage of the quality of the waste heat overcomes the small decrease in the system efficiency.

Solid oxide fuel cells hold perhaps the most promise for micro-CHP applications. The European Commission has recognized the natural gas powered, high-temperature fuel cell as being among the most efficient and environmentally friendly means of achieving distributed cogeneration. Several micro-CHP systems using SOFC technology have been developed and tested. Ceramic Fuel Cells Limited (CFCL) has been developing flat plate SOFC technology for thirteen years. In 2004 CFCL announced an all-ceramic stack design capable of volume fabrication for a 1-kW micro-CHP fuel cell system targeting the residential market. The SOFC stack developed by CFCL has been designed with commercial requirements in mind. The stack has the ability to reform fuel at the anode without the use of an external reformer and operates at temperatures of 1470 – 1600 F (800 – 870 C). The SOFC stack utilizes a modular design that allows stacks to be arranged in parallel to generate power in a range of 1 – 10 kW.

An example of an SOFC installed for residential micro-CHP use is at the Canadian Centre for Housing Technology in Ottawa. The unit is a 5-kW system that operates on natural gas, using a tubular arrangement developed by Siemens Westinghouse Power Corporation. The unit has the ability to use low fuel pressures which decreases operational costs. The system also includes an inverter that meets residential standards and output control that allows the system to output/receive electricity to the grid.

The SOFC unit has been placed in a research house that incorporates simulated occupancy and an "intensively monitored real world environment." Researchers intend to monitor heating, ventilation, and air-conditioning conditions to develop methods for better controls systems and grid connections.

The Energy Research Centre of the Netherlands (ENC) has also experimented with the use of a 1-kW SOFC micro-CHP unit. The system operates with an HXS 1000 Premiere fuel cell system installed in September 2002. The unit has a thermal output of 2.5 kW and is equipped with an auxiliary heater. The unit is fueled by natural gas, and the operating temperature is approximately 900 C. The

SOFC cells are manufactured by InDEC and are arranged in a stack assembly. The system is pictured in Figure 15.

The operation of the system begins with the desulphurization of the natural gas which is mixed with de-ionized water vapor. The mixture is internally reformed at the anode, and carbon monoxide and hydrogen are consumed at the anode. Waste heat is recovered by the combustion of the mixture of anode "off gas" (water vapor and carbon dioxide) and cathode exhaust. The combustion provides sensible heat to the storage vessel. (Oosterkamp et al., 1993)

A. Thermal insulation
B. SOFC stack
C. Heat Exchanger
D. Heat storage tank (200 l)
E. Process control unit
F. Auxiliary burner
G. DC/AC inverter
H. Desulphurisation unit
I. Exhaust line

Figure 15. SOFC Stack Assembly (one cell shown) (Oosterkamp et al., 1993).

The SOFC micro-CHP system operates in a heat following mode and has demonstrated the following performance:

- Electrical efficiency of 25-32% at full load
- Thermal efficiency of 53-60% on LHV basis
- Total efficiency of approximately 85%

However, the study also found that the electrical efficiency does decrease over time. The efficiency decrease is rather fast for strongly dynamic system operation, but is also noticeable for a more stable and steady operation.

PAFC

Phosphoric acid fuel cells are generally considered the "first generation" technology. PAFC fuel cells typically operate at a temperature of 390 F (200 C) and can achieve 40 % to 50 % fuel to electrical efficiencies on a lower heating value basis (LHV). The PAFC operates similar to the PEMFC. The PAFC uses a proton-conducting electrolyte (phosphoric acid), and the reactions occur on highly-dispersed electrocatalyst particles supported on carbon black. Phosphoric acid (H_3PO_4) is the only common inorganic acid that has enough thermal, chemical, and electrochemical stability and low volatility to be considered as an electrolyte for fuel cells. Typically, platinum is used as the catalyst at both the anode and cathode. The reactions that occur within a generic PAFC are as follows:

Anode Reaction: $H_2 \Leftrightarrow 2H^+ + 2e^-$
Cathode Reaction: $O_2(g) + 4H^+ + 4e^- \Leftrightarrow 2H_2O$
Overall Reaction: $2H_2 + O_2 \Leftrightarrow 2H_2O$

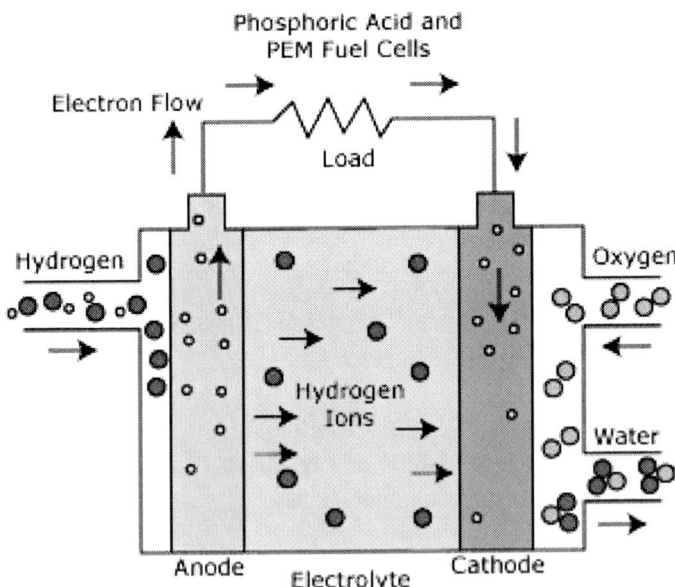

Figure 16. PAFC Reactions (Available at http://www.fctec.com/fctec).

In the PAFC reactions, hydrogen is ionized at the anode to produce two hydrogen ions and two electrons. The hydrogen ions pass through the electrolyte to the cathode while the electrons travel through a load in an external circuit. At the cathode, the hydrogen ions, electrons, and oxygen react to form water. The reactions that occur in the PAFC are shown in Figure 16.

The PAFC has experienced success in the large-scale CHP arena and holds potential for the micro-CHP market. However, currently few manufacturers are pursuing development of PAFC micro-CHP based system but have expressed more interest in the SOFC and PEMFC fuel cells. One reason for manufacturers' lack of interest in PAFC fuel cells is that the PAFC fuel cell uses only pure hydrogen as fuel and other fuel cell types have the capability to utilize a fuel other than a pure hydrogen stream. This selection could also result from potential hazards the electrolyte (phosphoric acid) could pose to inhabitants in the event of a leak. The PAFC requires an external fuel reformer, increasing the production cost.

MCFC

Molten Carbonate Fuel Cells (MCFC) are high-temperature fuel cells and promise the highest fuel-to-electricity efficiencies for carbon based fuels. The higher operating temperature allows the MCFC's to use natural gas directly without the need for a fuel reformer. MCFC have also been used with low-Btu fuel gas from industrial processes and other fuel sources. Developed in the mid 1960s, improvements have been made in fabrication methods, performance and endurance.

The MCFC operates differently than the fuel cells previously discussed. MCFC's use an electrolyte composed of a molten mixture of alkali metal carbonate salts, which is retained in a ceramic matrix of $LiAlO_2$. Two mixtures are currently used: lithium carbonate and potassium carbonate or lithium carbonate and sodium carbonate. To melt the carbonate salts and achieve high ion mobility through the electrolyte, MCFC's operate at temperatures of 600-700 C. When heated to a temperature of around 650 C, the salts melt and become conductive to carbonate ions (CO_3^{2-}). These ions flow from the cathode to the anode where they combine with hydrogen to give water, carbon dioxide and electrons. Electrons are routed through an external circuit back to the cathode, generating electricity and by-product heat.

Unlike other fuel cells, carbon dioxide as well as oxygen needs to be delivered to the cathode. The carbon dioxide and oxygen react to form carbonate ions, which provide the means of ion transfer between the cathode and anode. At the anode, the carbonate ions are converted back into carbon dioxide. There is a net transfer of CO_2 from the cathode to the anode; one mole of CO_2 is transferred for every two moles of electrons. The operation of a MCFC is illustrated in Figure 17. The reactions occurring in the MCFC are

Anode Reaction: $CO_3^{2-} + H_2 \Leftrightarrow H_2O + CO_2 + 2e^-$

Cathode Reaction: $CO_2(g) + \frac{1}{2}O_2 + 2e^- \Leftrightarrow CO_3^{2-}$

Overall Reaction: $2H_2 + \frac{1}{2}O_2 + CO_2(cathode) \Leftrightarrow 2H_2O + CO_2(anode)$

The higher operating temperature of the MCFC creates both advantages and disadvantages compared to fuel cells such as the PAFC and PEMFC with lower operating temperatures. The higher operating temperature allows fuel reforming of natural gas to occur internally, eliminating the need for a fuel reformer. The MCFC can also utilize standard materials for construction, such as stainless steel and nickel-based catalysts. The by-product heat from an MCFC can be used to generate domestic hot water, space heating and cooling, and even high-pressure steam.

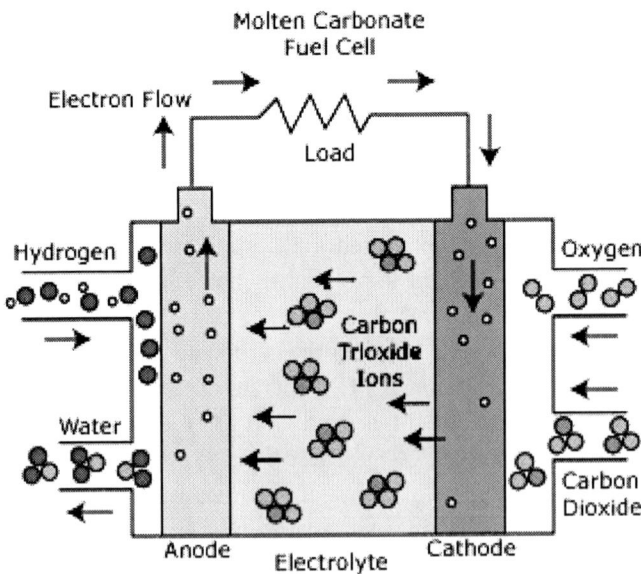

Figure 17. MCFC Reactions (Available at http://www.fctec.com/fctec).

Table 5. Overview of Fuel Cell Characteristics
(Available at www.energy.ca.gov/distgen/)

	Fuel Cells Overview			
	PAFC	SOFC	MCFC	PEMFC
Commercially Available	Yes	Yes	Yes	Yes
Size Range	100-200 kW	1 kW - 10 MW	250 kW - 10 MW	3-250 kW
Fuel	Natural gas, landfill gas, digester gas, propane	Natural gas, hydrogen, landfill gas, fuel oil	Natural gas, hydrogen	Natural gas, hydrogen, propane, diesel
Efficiency	36-42%	45-60%	45-55%	25-40%
Environmental	Nearly zero emissions	Nearly zero emissions	Nearly zero emissions	Nearly zero emissions
Other Features	Cogen (hot water)	Cogen (hot water, LP or HP steam)	Cogen (hot water, LP or HP steam)	Cogen (80°C water)
Commercial Status	Some commercially available	Commercialization likely in 2004	Some commercially available	Some commercially available

The high operating temperatures and the electrolyte chemistry of the MCFC also lead to disadvantages. The high temperature requires significant time to reach operating conditions and correspondingly slow response time to changing power demands. These characteristics make the MCFC less attractive for dynamic power applications and restrict it to constant-power supply applications. The carbonate electrolyte can also cause electrode corrosion problems. Due to the use of the carbonate ions as the charge carrier, the supply of carbon dioxide to the cathode must be carefully controlled in order to achieve optimum performance. Each fuel cell type has both advantages and disadvantages over its counterparts. The selection of the "best" fuel cell will depend upon the details of the application. An overview of fuel cell characteristics is given in table 5.

Application

Fuel cells are being developed for stationary power in small commercial and residential markets and as peak shaving units for commercial and industrial uses. Some fuel cells, such as PEMFC, are currently undergoing development for use in automobiles and portable power applications.

Phosphoric acid fuel cells have been installed at medical, industrial, and commercial facilities throughout the country, and the 200-kW size is a good match for distributed generation applications. The operating temperature is about 400 F, which is suitable for co-generation applications. Developers are targeting commercial and light industrial applications in the 100-200 kW power range, for both electric-only and cogeneration applications.

The high efficiency and high operating temperature of MCFC units makes them most attractive for base-loaded power generation, either in electric-only or cogeneration modes. Potential applications for the MCFC are industrial, government facilities, universities, and hospitals.

Solid oxide fuel cells are being considered for a wide variety of applications, especially in the 5 – 250 kW size range. These applications include small commercial buildings, industrial facilities, micro-CHP, and base load utility applications.

Proton exchange membrane fuel cells are currently undergoing the most rapid development of any fuel cell type. Part of this development has been driven by the desire of automotive manufacturers to develop a fuel cell powered automobile. This surge in development has led to breakthroughs for stationary power applications as well. Research is aimed at commercial-sized power generation (e.g., Ballard's 250 kW unit) and residential power generation (e.g., Plug Power's 3-5 kW units). For the units to achieve market potential, natural gas is selected as the fuel of choice. Reject system heat in the form of hot water makes them particularly attractive for cogeneration, which is included in almost all products currently under development.

Heat Recovery

The type of fuel cell determines the temperature of the heat rejected during operation and directly influences the fuel cell type's suitability for micro-CHP applications. Low temperature fuel cells create waste heat suitable for producing hot water and in some cases, low pressure steam. Lower temperature fuel cells such as the PAFC and PEMFC produce lower quality waste heat and are suitable for small commercial and industrial cogeneration applications. The MCFC and the SOFC operate at high temperatures and are capable of producing waste heat that can be used to generate steam for use in a steam turbine, or combined cycle microturbine. If space cooling is considered and an absorption chiller is to be used, the recaptured heat should be at a temperature of at least 185 F (85 C).

Cost

The initial cost of fuel cells is higher than those of other electricity generation technologies. The only product available commercially today is the PureCell 200 (formerly PC-25)™ built by UTC Power. The cost of the unit is approximately $4,000/kW. The installed cost of the unit approaches $1.1 million. At a rated output of 200 kW, this translates to about $5,500/kW, installed. However, on January 3, 2005, Delphi Corp., in partnership with the DOE's advanced fuel cell development program, reported that researchers have exceeded the government's $400/kW power cost goal for fuel cells. At this price, fuel cells could compete with traditional gas turbine and diesel electric generators and become viable power suppliers for the transportation sector. Table 6 shows the uninstalled projected long-term costs of fuel cell technologies. The price of $400/kW is not included in table 4-7 as the information was only recently released and currently, the development has not been proven for production costs in a fuel cell of substantial size.

As no combustion is occurring, and there are no moving parts, fuel cells are expected to have minimal maintenance requirements. The primary maintenance will be focused on preventing poisoning of the catalyst and periodic inspection and maintenance to the fuel supply system and fuel reformers. The cell stack itself will not require maintenance until the end of its service life. Fuel cell system maintenance requirements vary with the type of fuel cell, size, and maturity of the equipment. Major overhaul of fuel cell systems involves shift catalyzer replacement, reformer catalyzer replacement, and stack replacement. The maintenance and reliability of the system still needs to be proven in a long-term demonstration. Maintenance costs of a fuel cell are expected to be comparable to that of a microturbine, ranging from $0.005-$0.010/kWh (based on an annual inspection visit to the unit). *(Available at http://www.energy.ca.gov/distgen/)*

Table 6. Projected Long-Term Costs of Fuel Cell Technologies (Available at www.energy.ca.gov/distgen/)

Emerging Fuel Cell Technologies	
Technology	Projected Cost (Long-term, Uninstalled)
MCFC	$1,200-1,500/kW
SOFC	$1,000-1,500/kW
PEMFC	Initially $5,000/kW Long term $1,000/kW

Table 7. Advantages and Disadvantages of Fuel Cell Types (Available at www.energy.ca.gov/distgen/)

PAFC	
Advantages	Disadvantages
Quiet	High Costs
Low emissions	
High efficiency	Low energy density
Proven reliability	
PEMFC	
Advantages	Disadvantages
Quiet	High Costs
Low emissions	
High efficiency	Limited field test experience
Synergy with Automotive R&D	Low temperature waste heat may limit cogeneration potential
SOFC	
Advantages	Disadvantages
Quiet	High Costs
Low emissions	
High efficiency	Planar SOFC's are still in the R&D stage but recent developments in low temperature operation show promise
High energy density	
Self reforming	
MCFC	
Advantages	Disadvantages
Quiet	High Costs
Low emissions	
High efficiency	Need to demonstrate long term reliability
Self reforming	

Advantages and Disadvantages

Each fuel cell type will have advantages and disadvantages in certain areas, both as compared to other fuel cell technologies and other DER equipment. Table 7 displays some of the advantages of the four primary types of fuel cells.

Fuel cells convert chemical energy directly into electricity without the combustion process. As a result, a fuel cell does not incur losses resulting from mechanical inefficiencies. Fuel cells can achieve high efficiencies in energy conversion terms, especially where the waste heat from the cell is utilized in cogeneration. A high power density allows fuel cells to be a relatively compact source of electric power, a benefit in applications with space constraints. In a fuel cell system, the fuel cell itself is often smaller than the other components of the system such as the fuel reformer and power inverter.

Fuel cells, due to their nature of operation, are extremely quiet in operation. This allows fuel cells to be used in residential or built-up areas where the noise pollution is undesirable. Unfortunately, the primary disadvantage of the fuel cells is the cost. The two basic reasons are high component costs compared to other energy systems technology and fuel cell operation requires a continuous, highly selective, expensive fuel supply.

HEAT RECOVERY

Electrical power generation devices do not convert 100% of an energy source potential into usable energy. Electrical efficiencies of reciprocating engines, microturbines, Stirling engines, and fuel cells are about 50%, 30%, 30%, and 60%, respectively. This means that the electrical power generation devices fail to utilize 40% - 70% of the available energy. Energy that is not converted to electrical or shaft power is rejected from the process in the form of waste heat. In order to utilize more of the energy stored in a given fuel and increase the overall thermal efficiency of a system, heat recovery must be incorporated into a system. Heat recovery converts waste heat to useful energy and is primarily accomplished through the use of heat exchangers.

Distributed energy generation prime movers possess waste heat that can be recovered as useful energy. The type of prime mover determines the characteristics of the waste heat and the effectiveness with which useful energy can be recovered. Waste heat is typically released in the form of hot exhaust gases, process steam, or process liquids/solids. The usable temperature for heat recovery is listed for various prime movers in table 8. Potential uses for the waste heat are hot water, space heating and cooling, and process steam.

Table 8. Waste Heat Characteristics of Prime Mover Technologies

	Usable Temp. for micro-CHP (°F)
Diesel Engine	180 - 900
Natural Gas Engine	300 - 500
Stirling Engine	800 - 1000
Fuel Cell	140 - 750
Microturbine	400 - 650

Recovered heat that is utilized in the power generation process is internal heat recovery, and recovered heat that is used for other processes is external heat recovery. Recuperators, turbochargers, and combustion pre-heaters are examples of internal heat recovery. Absorption chillers, desiccant dehumidification devices, and heat recovery steam generators are examples of external heat recovery components. Recovered heat is classified as low-temperature when less than 445 F (<230 C), medium-temperature 445 F – 1202 F (230 C – 650 C), or high-temperature when greater than 1202 F (>650 C). The majority of heat exchangers used in micro-CHP are used for external heat recovery.

Technology Overview

Because combustion exhaust gases or process liquids cannot be used directly in many applications, heat exchangers are used to facilitate the transfer of waste heat thermal energy to heat recovery applications. Devices that transfer energy between fluids at different temperatures by heat transfer modes are known as heat exchangers. Heat exchangers come in a wide variety of sizes, shapes, and types and utilize a wide range of fluids. Applications for heat exchangers are vast, ranging from heating and air-conditioning systems, to chemical processing and power production in large plants. Heat exchanger classification is based upon component and fluid characteristics. Several classification schemes have been proposed for heat exchangers. Hewitt et al., (1994) suggests the following four-tiered system:

1. Recuperator/Regenerator
2. Direct-contact/Transmural heat transfer
3. Single phase/Two-phase
4. Geometry

A recuperator is based on a continuous transfer of heat between two fluid streams. A regenerator is a device which uses a heat-absorbing material, alternately cooled and heated in a batch mode, to transfer heat between two streams; these are often rotary devices. Direct-contact heat exchangers, such as a feedwater heater used in a power plant, allow the two fluids to come into contact with each other. Transmural heat exchangers separate the two fluid streams using a wall or series of walls. Single phase/two-phase refers to the physical state of the fluids flowing in the heat exchanger. Single-phase flow implies that both fluids are either completely gaseous or liquid. If evaporation or condensation is taking place, the device involves two phases. Geometry refers to the basic shape of the heat-exchanger passages that contain the fluid.

This section will focus on transmural recuperators with fluids in single-phase; however, other arrangements will also be discussed. Transmural recuperators with single-phase flow can be divided further by flow arrangement into the following categories:

- Counterflow
- Cross-flow
- Parallel flow

Counterflow and parallel flow tube-within-a-tube and cross-flow configurations are shown in Figure 18.

The temperature versus area diagrams for the parallel flow and counterflow arrangements are shown in Figure 19. In the parallel arrangement, the fluid temperatures approach each other so that the temperature difference ΔT_1 is much greater in magnitude than ΔT_2. If the length of the heat exchanger were extended long enough in parallel flow, the exit temperatures of each of the fluids would be approximately equal. The counterflow average ΔT will be larger than the parallel flow average ΔT.

Cross-flow heat exchangers are arranged so that the two fluids flow perpendicular to each other as shown in Figure 20. An important concept when discussing cross-flow heat exchangers is *mixing*. A fluid is said to be *unmixed* if the passageway contains the same discrete portion of fluid throughout its traverse of the exchanger. If the fluid passageways are such that fluid from one passageway can mix with fluid from others, the fluid is *mixed*. In Figure 20, mixed and unmixed arrangements are illustrated. Cross-flow heat exchangers are classified as both fluids unmixed, both fluids mixed, or one fluid mixed and one fluid unmixed.

When several tubes are placed inside a shell, the classical single-pass shell-and-tube heat arrangement results. A schematic of a shell-and-tube heat exchanger with one shell pass and one tube pass is illustrated in Figure 21. Baffles are often placed inside a shell-and-tube heat exchanger to promote higher heat transfer rates and increase the effectiveness of the heat exchanger.

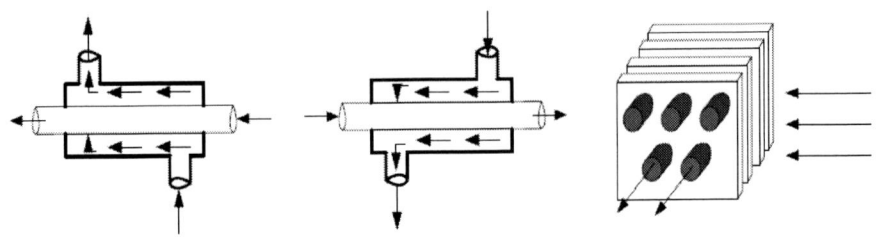

Figure 18. Main Types of Transmural Recuperators with Fluids in Single-phase.

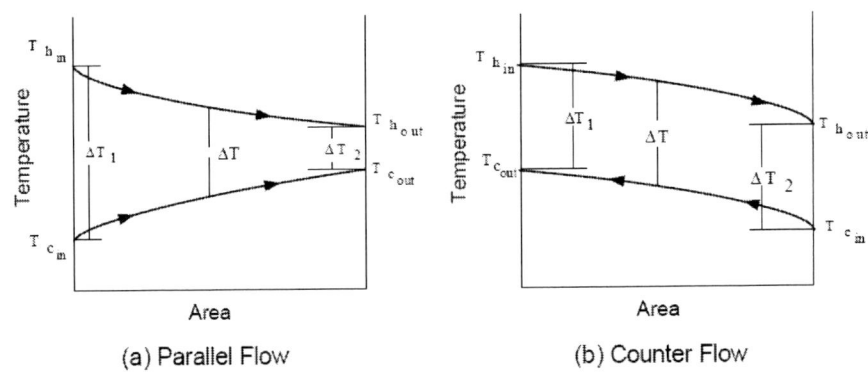

(a) Parallel Flow

(b) Counter Flow

Figure 19. Temperature-area Diagram of Parallel and Counterflow Arrangements.

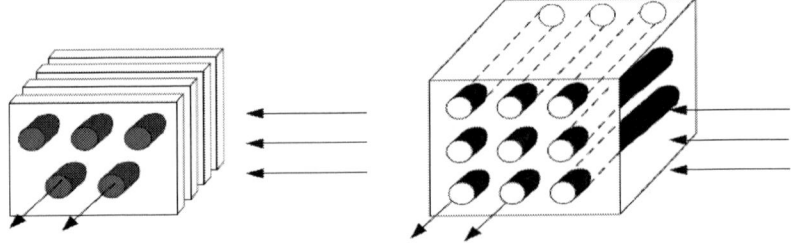

Figure 20. Cross-flow Heat Exchangers.

Prime Movers 45

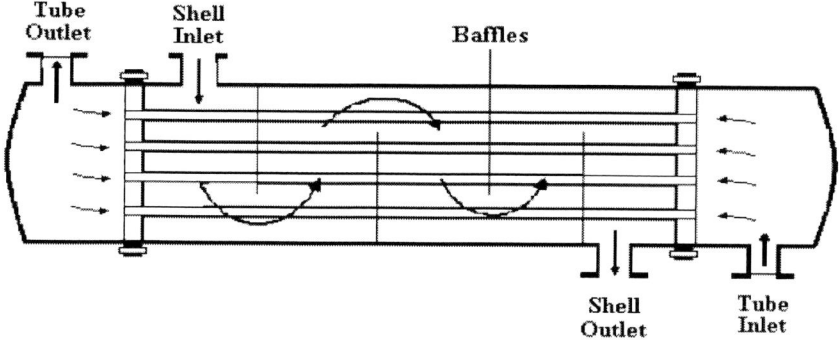

Figure 21. Shell-and-tube Heat Exchanger.

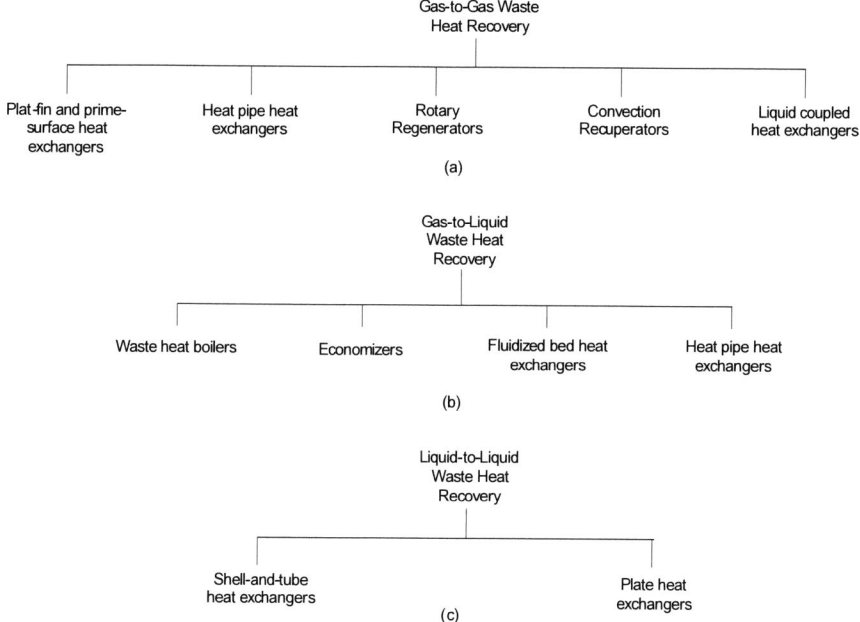

Figure 22. Classification of Waste Heat Recovery Heat Exchangers (Recreated from the *CRC Handbook of Energy Efficiency*).

Application

Waste heat recovery heat exchangers may be classified as gas-to-gas, gas-to-liquid, and liquid-to-liquid. The various types of these classifications are displayed in Figure 22.

Gas-to-Gas Heat Exchangers

Gas-to-gas waste heat recovery exchangers can be used as recuperators to preheat combustion air in IC engines and microturbines. Rotary regenerators are often used in Stirling engines to recover and store heat. Gas-to-gas heat exchangers find many applications in micro-CHP systems.

Metallic radiation recuperators, convection recuperators, and a runaround coil are the three primary types of gas-to-gas heat exchangers. A metallic radiation recuperator is a tube-in-tube heat exchanger that consists of two concentric metal tubes. Hot exhaust (flue) gas flows through the inner tube, and the air to be preheated flows through the outer tube or annulus. This type of recuperator can act as a part of a chimney, flue, or exhaust line. The majority of the heat is transferred from the hot gas to the inner wall through radiation. The heat transfer in the outer tube then takes place through convection. Air and gas flowing in counterflow is the most desirable arrangement because this arrangement has a high performance. Metallic radiation recuperators typically achieve an effectiveness of 40% or lower. Though metallic radiation recuperators could be used in micro-CHP applications, their use will likely be limited by the relatively low effectiveness which would result in large heat exchangers.

Convection recuperators are cross-flow heat exchangers with flue gases flowing normal to a tube bundle with air to be preheated flowing in the tubes. Convection recuperators can be used in low-temperature applications such as space heating, direct-fired absorption chillers, or return air heating in a desiccant dehumidification system.

A runaround coil consists of two connected heat exchangers that circulate a working fluid. The working fluid is heated by the waste gas with one heat exchanger and then circulated to the other heat exchanger where the fluid heats a stream of cool air. Runaround coils are used in HVAC applications and can be coupled with distributed generation components to produce warm air for district heating, to heat return air for desiccant dehumidification devices, or to fire an absorption chiller (Shah, 1997).

Gas-to-Liquid Heat Exchangers

Gas-to-liquid heat exchangers include economizers, waste heat boilers, heat-pipe heat exchangers, fluidized-bed, and thermal fluid heaters. Economizers and thermal fluid heaters are used for low- to medium-temperature waste heat recovery while waste heat boilers, heat-pipe heat exchangers, and fluidized bed heat exchangers are used for medium- to high-temperature heat recovery. Economizers are most often used with boilers to preheat the boiler feedwater. In other applications, economizers are referred to as a process fluid or water heaters. An economizer is an individually finned-tube bundle, with gas flowing outside normal to the finned tubes and water flowing inside the tubes.

Thermal fluid heaters are double-pipe heat exchangers that use waste heat gases to heat a high-temperature organic heat transfer fluid. Thermal fluid heaters operate on the same principle as a domestic warm-water system, except that the water is replaced by a high-temperature organic heat transfer fluid. The heat transfer fluid can be circulated and used for heating and heat-driven absorption chillers.

Fluidized-bed heat exchangers utilize water, steam, or another heat transfer fluid heated by waste heat gases that flow over a bed of finely divided solid particles. As the waste heat fluid reaches a critical velocity, the particles begin to float, and the resulting mixture acts as a fluid.

Economizers and thermal fluid heaters are more likely to be used in micro-CHP systems than either boilers or fluidized-bed heat exchangers due to the amount of thermal energy available for heat recovery and the size requirements of fluidized-bed heat exchangers. Process steam is also less likely to be needed for micro-CHP applications, eliminating the need for waste heat boilers.

Liquid-to-Liquid Heat Exchangers

Liquid-to-liquid waste heat recovery heat exchangers are typically used in industrial applications. Shell-and-tube heat exchangers and plate heat exchangers are typically used for liquid-to-liquid heat recovery. Liquid coolant systems in reciprocating engines offer liquid-to-liquid heat recovery opportunities from hot oil and other liquid coolants. However, liquid-to-liquid heat recovery will not be as significant a contributor in micro-CHP applications as are gas-to-gas and gas-to-liquid waste heat recovery exchangers. (Shah, 1997)

Chapter 4

THERMALLY ACTIVATED DEVICES

Thermally activated devices are technologies that utilize thermal energy rather than electric energy to provide heating, cooling, or humidity control. The primary thermally activated components used in micro-CHP systems are desiccant dehumidifiers and absorption chillers.

ABSORPTION CHILLERS

Technology Overview

Absorption chiller technologies are one of a group of technologies classified as heat pumps. Heat pumps may be either heat driven or work driven. Absorption technologies are heat driven, transferring heat from a low temperature to a high temperature using heat as the driving energy. Heat pumps operate on the principle of the absorption refrigeration cycle, which is similar to the vapor-compression cycle. Both the absorption refrigeration cycle and the vapor-compression cycle will be examined to draw analogies between the two.

Vapor-compression refrigeration systems are the most common refrigeration systems in use today. The vapor-compression cycle is a work-driven cycle that is illustrated in Figure 23. In the vapor-compression cycle, work is input to compress the refrigerant to a high pressure and temperature at State 2. At State 2, the refrigerant condensation temperature is below the ambient temperature. As the high-pressure and high-temperature refrigerant vapor passes through the condenser, heat is rejected to the ambient air and the refrigerant vapor condenses to a liquid to achieve State 3. The high-pressure liquid at State 3 passes through an expansion valve. As the liquid passes through the expansion valve, the refrigerant experiences a reduction in both temperature and pressure to reach State 4. At State 4, the boiling temperature of the refrigerant is lower than that of the surroundings. The low-pressure liquid refrigerant passes through the evaporator, absorbing heat from the ambient environment when boiling occurs in the evaporator and creating a low-pressure refrigerant vapor at State 1. The low-pressure refrigerant vapor at State 1 enters the compressor completing the cycle.

The absorption cycle has some features in common with the vapor compression cycle. For example, the absorption cycle has a condenser, an evaporator, and an expansion valve. However, the absorption cycle and the vapor-compression cycle differ in two very important aspects. The absorption cycle uses a different compression process and different refrigerants than the vapor-compression cycle.

The absorption cycle operates on the principle that some substances (absorbents) have an affinity for other liquids or vapors and will absorb them under certain conditions. Instead of compressing a vapor between the evaporator and condenser as in Figure 23, the refrigerant of an absorption system is absorbed by an absorbent to form a liquid solution. The liquid solution is then pumped to a higher pressure. Because the average specific volume of a liquid is much smaller than that of the refrigerant vapor, significantly less work is required to raise the pressure of the refrigerant to the condenser pressure. This corresponds to less work input for an absorption system as compared to a vapor-compression system.

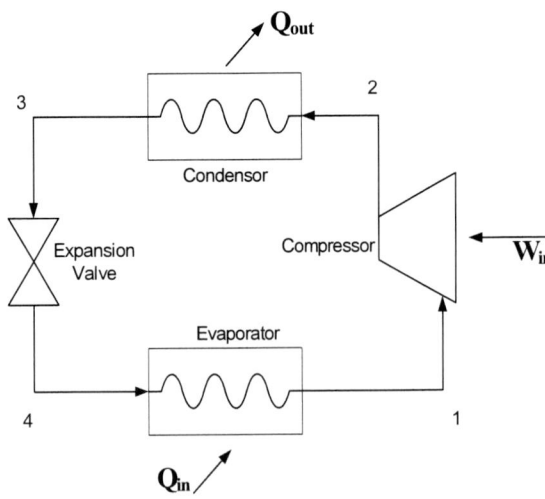

Figure 23. Vapor-Compression Cycle Schematic.

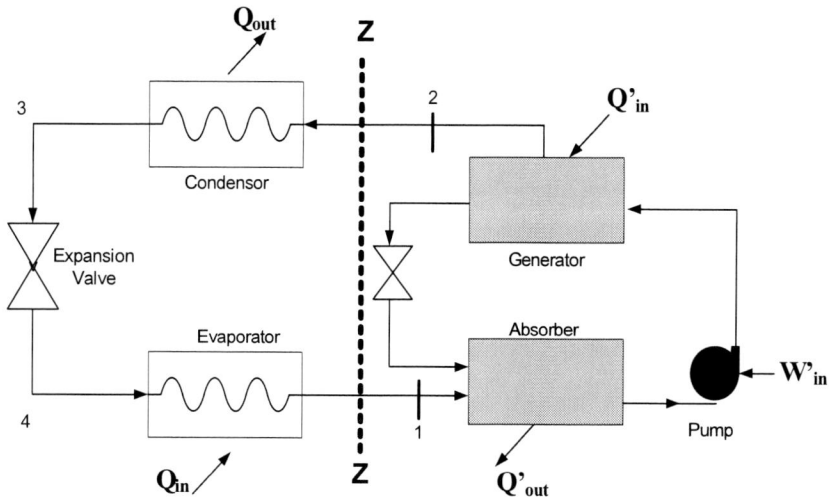

Figure 24. Basic Absorption Cycle Schematic.

Because the absorbent used in the absorption cycle forms a liquid solution, some means must also be introduced to retrieve the refrigerant vapor from the liquid solution before the refrigerant enters the condenser. This process involves heat transfer from a relatively high-temperature source. Because the thermal energy input into the system is much higher than the work input through the pump, absorption chillers are considered to be heat driven. The components used to achieve the pressure increase in an absorption chiller are viewed as a "thermal compressor" and replace the compressor in the vapor-compression cycle shown in Figure 23. The components of the absorption cycle are shown schematically in Figure 24. The components of the thermal compressor are a pump, an absorber, and a (heat) generator and are shown to the right of the dashed Z-Z line. The components to the left of the dashed Z-Z line are the same as the ones used in the vapor-compression system.

The operation of the absorption cycle shown in Figure 24 is as follows: At State 1, the low-pressure refrigerant vapor exits the evaporator and enters the absorber. In the absorber, the refrigerant vapor is dissolved in an absorbent and rejects the heat of condensation and the heat of mixing to form a liquid solution. The refrigerant/absorbent solution is then pumped to the condenser pressure and passed to the generator. In the generator, heat is added to the refrigerant/absorbent solution to vaporize the refrigerant, removing the refrigerant from the solution. The liquid absorbent has a higher boiling temperature than the refrigerant and, therefore, stays in the liquid form. There are two streams exiting the generator.

The refrigerant exits to the condenser at a high temperature and pressure (State 2) while the absorbent passes through an expansion valve, decreasing the pressure of the absorbent to the evaporator pressure before entering the absorber again.

The remainder of the operation is much the same as the vapor compression cycle. The high-temperature, high-pressure refrigerant vapor at State 2 enters the condenser with a pressure such that the ambient temperature is higher than the condensation temperature of the refrigerant. The refrigerant vapor condenses as it passes through the condenser, rejecting heat to the ambient environment to achieve State 3. At State 3, the high-pressure, low-temperature liquid refrigerant enters the expansion valve where the refrigerant experiences a decrease in pressure to the evaporator pressure. The low-pressure, low-temperature liquid refrigerant that results at State 4 is at a pressure such that the boiling temperature of the refrigerant is lower than the ambient temperature of the environment. As the liquid refrigerant passes through the evaporator, the refrigerant boils, absorbing heat from the ambient air. The refrigerant exits the evaporator as a high-temperature, low-pressure vapor to complete the cycle.

REFRIGERANT-ABSORBENT SELECTION

Though all absorption chillers operate on the basic cycle presented in Figure 24, each chiller design is dependent on the refrigerant-absorbent selection. Current refrigerant/ absorber media for absorption chillers are either water/lithium bromide or ammonia/water. Water/lithium bromide absorption chillers utilize water as the refrigerant and lithium bromide as the absorbent. Because water is used as the refrigerant, applications for the water/lithium bromide absorption chillers are limited to refrigeration temperatures above 0 C. This combination of refrigerant and absorbent is advantageous in areas where toxicity is a concern because lithium bromide is relatively non-volatile. Absorption machines based on water/lithium bromide are typically configured as water chillers for air-conditioning systems in large buildings. Water/lithium bromide chillers are available in sizes ranging from 10 to 1500 tons. The coefficient of performance (COP) of these machines typically falls in the range of 0.7 to 1.2 (Herold et al., 1996).

The ammonia/water combination utilizes ammonia as the refrigerant and water and the absorbent. The use of the ammonia as the refrigerant allows much lower refrigerant temperatures (the freezing temperature of ammonia is -77.7 C); however, the toxicity of ammonia is a factor that has limited the use of ammonia/water chillers to well-ventilated areas. In commercial and residential building applications where there is insufficient ventilation, emissions from an ammonia/water absorption chiller could be harmful to occupants.

The primary selling point of ammonia/water absorption chillers is their ability to provide direct gas-fired and air-cooled air conditioning. Ammonia/water absorption chillers are commonly sold as a single component in an air-conditioning system; however, use is restricted in some densely populated areas. Ammonia/waster chillers are available in capacities ranging from 3 to 25 tons with COPs generally around 0.5 (Herold et al., 1996).

A schematic of an absorption cycle of ammonia/water is shown in Figure 25. The addition of a heat exchanger is common in all absorption chillers to increase the efficiency of the thermal compressor. The hot solution leaving the generator is used to preheat the refrigerant/absorbent solution entering the generator. A rectifier is also included in the system of an ammonia/water chiller. This is because the ammonia vapor leaving the generator often includes a low concentration of water vapor. This water vapor will freeze in the expansion valve, negating the operation of the system.

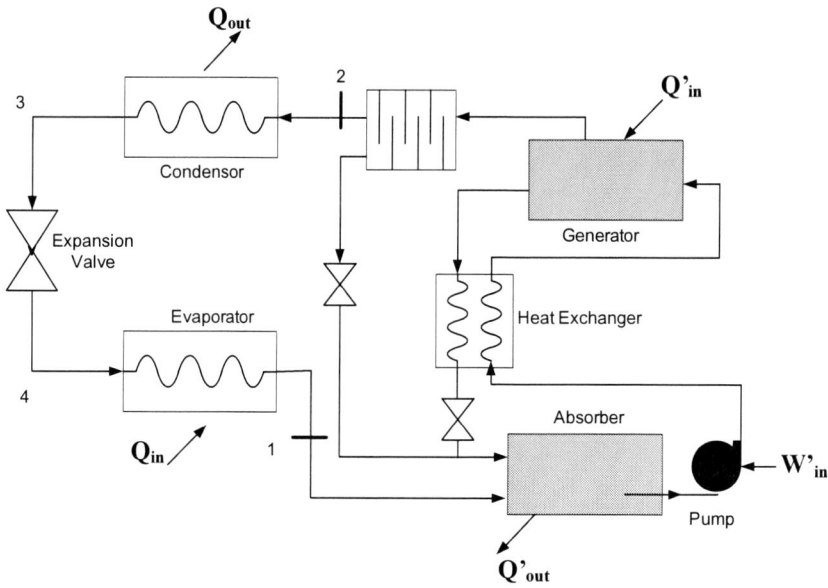

Figure 25. Ammonia/Water Absorption Cycle.

Types of Absorption Chillers

Absorption chillers are classified as single-effect, double-effect, or triple effect. Single-effect absorption chillers contain one stage of generation, such as the systems shown in figures 24 and 25. Single-effect absorption chillers use low-pressure steam or hot water as the energy source. Typical temperature requirements range from 200 to 270 F (93 to 132 C). Steam-powered systems operate at pressures between 9 and 15 psig. When the supplied temperatures are below the design specifications, the chiller capacity is reduced.

Double-effect absorption systems use a second generator, condenser, and heat exchanger that operate at higher temperature. A double-effect water/lithium bromide absorption system is shown schematically in Figure 26. Refrigerant vapor is recovered from the first-stage generator in the high-temperature condenser. The refrigerant/absorbent in the second-stage generator is at a lower temperature than the solution in the first-stage generator. The refrigerant vapor from the first stage generator flows through the second-stage generator where a portion of the refrigerant condenses back into liquid while the remainder remains in the vapor phase. Additional refrigerant is vaporized in the second-stage generator by the high temperature and the heat of vaporization supplied by the refrigerant from the first-stage generator. The refrigerant vapor from both generator stages flows to the condenser while the absorbent solution flows back to the absorber.

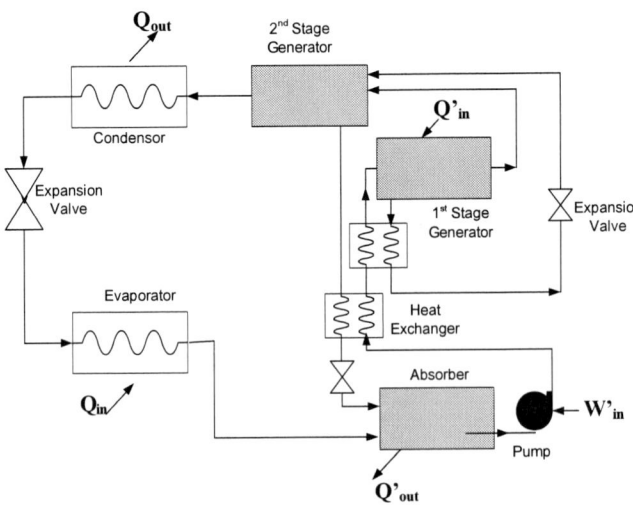

Figure 26. Double-Effect Water/Lithium Bromide Absorption Chiller Schematic.

The purpose of the two stages that make up the double-effect absorption cycle is to increase the COP of the cycle. This is made possible through the use of the recuperative heat exchangers used in the system. Double-effect chillers yield higher COPs than single-effect chillers. The COP for double-effect absorption chillers varies from 1.0 to 1.2 for water/lithium bromide chillers (Herold et al., 1996).

Triple-effect absorption chillers have been in prototype development for several years. These chillers will be direct-fired and are expected to provide a 50% thermal efficiency improvement over double-effect absorption chillers. Triple-effect absorption chillers do not feature a distinct third generator stage; rather they use internally-recovered heat to achieve high efficiencies. Triple-effect water/lithium bromide chillers can achieve COPs of 1.6 and greater (Petchers, 2003).

One of the most promising absorption technologies is the generator-absorber heat exchange (GAX) cycle. GAX chillers use an ammonia/water working fluid. GAX-cycle systems hold great potential for residential and light-commercial applications and provide capacities as low as 3 tons. GAX absorption chillers have obtained COPs of approximately 0.7.

Application

Absorption chillers can be directly fired or indirectly fired. Direct-fired absorption chillers utilize a natural gas burner and can supply waste heat for a desiccant dehumidification device or hot water. Direct-fired chillers are often used in areas where electric rates are high and gas utilities offer lower rates or rebate programs to replace vapor-compression chillers.

Indirect-fired fired absorption chillers are utilized where there is an existing source of heat that can be recovered. The supplied heat can be in the form of hot water, steam, or exhaust gases. All of the prime mover technologies that are applicable to micro-CHP can produce waste heat sufficient to drive an absorption chiller. This coupling ability makes absorption cycle chiller systems very desirable for micro-CHP applications. An absorption chiller in an micro-CHP system may not utilize all of the waste heat that is input into the chiller. Just as with direct-fired chillers, this remaining heat may be used in a desiccant dehumidification device or to produce hot water.

The temperature of the waste heat available from a power source determines the appropriate absorption configuration. Table 9 matches the waste heat temperatures typical of various prime movers with appropriate absorption configurations. The match is based on the temperature of the waste heat that could be obtained to drive the generator in an absorption chiller cycle.

Absorption chillers offer many advantages over electric chillers, especially when there is a source of waste heat available. As compared to electric chillers, absorption chillers have lower operating costs, shorter payback periods, quiet operation, low maintenance, and high reliability. Absorption systems also operate at lower pressures and offer safer operation. However, absorption chillers have higher initial costs and are not as widely available as electric chillers.

Vapor-compression systems are much more widely manufactured and more available than absorption chillers. Still, the fact that absorption chillers do not have mechanical compressors and have fewer moving parts gives absorption technologies an advantage over vapor-compression systems in terms of lower maintenance, higher reliability and quieter operation.

Table 9. Matching of Power Generation and Absorption Technology

Power Generation and Absorption Technology		
Power Source	Temperature (F)	Matching Technology
Microturbine	~600	Triple-, double-, or single-effect
Reciprocating Engine	~180	Single-effect
SOFC	~900	Triple-, double-, or single-effect
PAFC	~250	Double-effect or single effect
PEMFC	~140	Single-effect

Cost

The capital cost of installing an absorption chiller is generally more than installing an equivalent electric or engine-driven chiller. The RS Means Mechanical Cost Data (2001) for a 100 ton steam or water-fired absorption chiller is presented in table 10 (replace with 10 ton unit cost).

Table 10. RS Means Cost Data for a 100 Ton Absorption Chiller Installation

Cost Source	Cost ($)
Material	110,500
Labor	7,975
Total	118,475

DESICCANT DEHUMIDIFICATION TECHNOLOGIES

Controlling humidity in a conditioned space can be important for a number of reasons. Concerns include humidity damage to moisture sensitive items, product protection from moisture degradation (foods, grains, and seeds), mildew growth, corrosion, and health issues. In the last thirty years, applications have expanded for desiccant technologies to include supermarkets, hospitals, refrigerated warehouses, ice rinks, hotels, and retail establishments. Recently, development has turned to applying desiccant dehumidification to commercial and residential buildings.

Desiccants are materials that attract and hold moisture. A desiccant dehumidifier is a device that employs a desiccant material to produce a dehumidification effect. The process involves exposing the desiccant material to a high relative humidity air stream, allowing the desiccant to extract and retain a portion of the water vapor, and then exposing the same desiccant to a lower relative humidity air stream where the retained moisture is drawn from the desiccant.

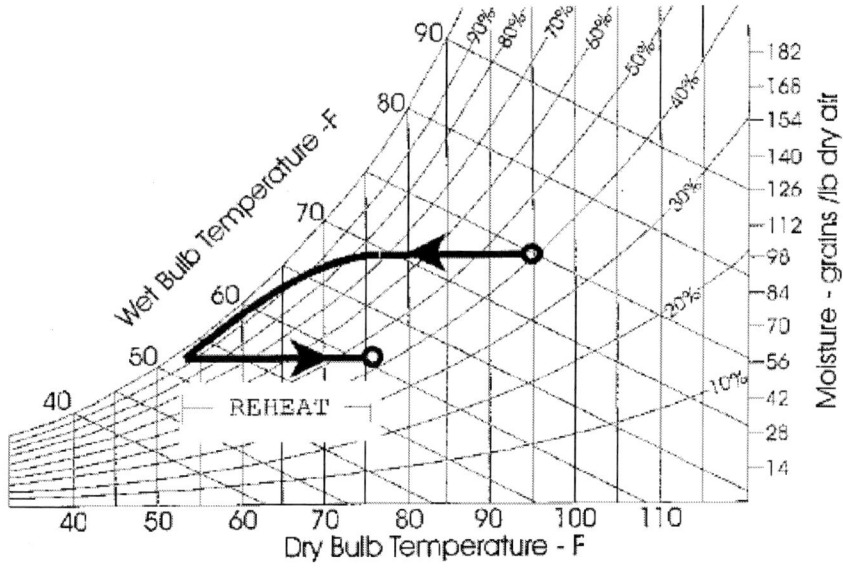

Figure 27. Sub-cooling Dehumidification Process (Chamra et al., 2000).

This section presents an overview of desiccant dehumidification. The principles of the sub-cooling system are introduced to contrast the differences between desiccant and conventional air dehumidification and to highlight the advantages of desiccant dehumidification. Also discussed are the principles of desiccant systems, types of desiccant systems, and cost considerations for choosing desiccant systems. Desiccant dehumidification technologies are an attractive component of an micro-CHP system because desiccant regeneration provides an excellent use for waste heat.

Sub-cooling Systems vs. Desiccant Systems

In traditional air-conditioning systems, such as the units installed in most U.S. homes, dehumidification is achieved by cooling a moist air stream to a temperature below the dewpoint so that water (liquid) condenses out of the air. This process is familiar to anyone who has seen moisture condense on the exterior of a glass of ice water on a humid day. An example of the sub-cooling process is illustrated on a psychometric chart in Figure 27. The processes shown are for air being cooled and dehumidified from conditions of 95 F dry bulb (db) and 75 F wet bulb (wb) to 77 F db and 58 grains/lbm_{da}. The resulting air lies approximately in the center of the ASHRAE Summer Comfort Zone shown in Figure 28. Initially, the dry bulb temperature of the air decreases, while the moisture content remains constant. The dry bulb temperature continues to decrease as moisture begins to condense out of the air onto the cooling coil, resulting in a decrease in the moisture content. In order to deliver air at the desired drier condition of approximately 45% relative humidity, some form of reheating must be used. The reheat process path is also illustrated in Figure 27. In this example, the total net cooling load is 10.8 Btu/lbm_{da}, and of this, 6.4 Btu/lbm_{da}, or about 59%, is latent load.

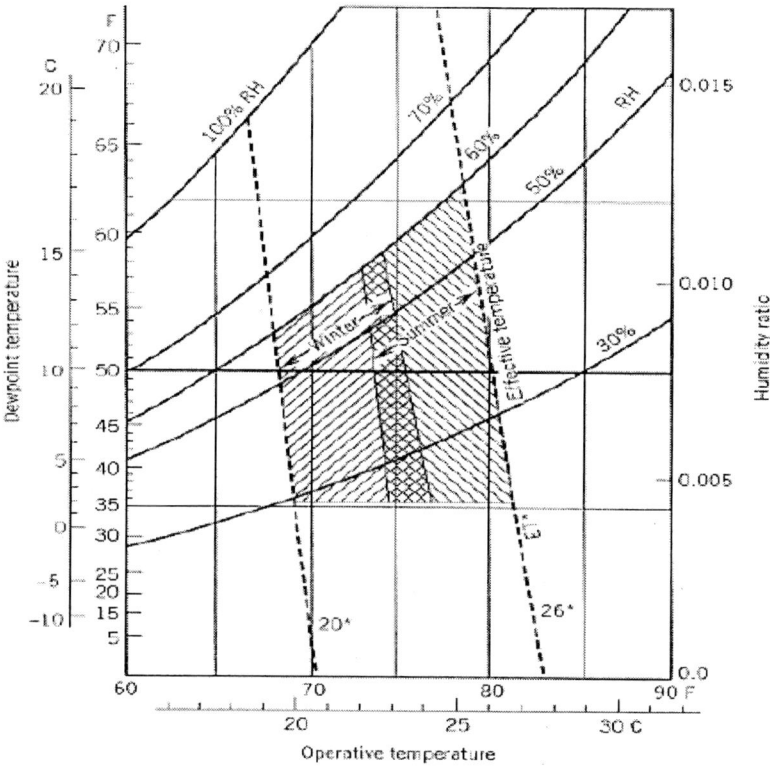

Figure 28. ASHRAE Comfort Zones (ASHRAE *Fundamentals*, 2001).

Summary of the Principles of Sub-cooling Systems

The same equipment is used for both the sensible cooling and dehumidification in the conventional system. If independent humidity and temperature control are required in a system, a provision for reheat of the cooled air must be included. In the example above, the net cooling load is 10.8 Btu/lbm$_{da}$, but the load on the cooling coil is 16 Btu/lbm$_{da}$ with the difference (5.3 Btu/lbm$_{da}$) being added back in during the reheat process. Thus, energy is used both for the extra cooling and for reheat.

Another disadvantage of this approach is that the air leaving the evaporator coil is nearly saturated, with relative humidity typically above 90%. This moist air travels through duct work until the air is either mixed with dryer air or reaches the reheat unit. The damp ducts, along with the wet evaporator coils and standing water in a condensate pan (Figure 29), can generate problems with microbial growth and the associated health and odor problems.

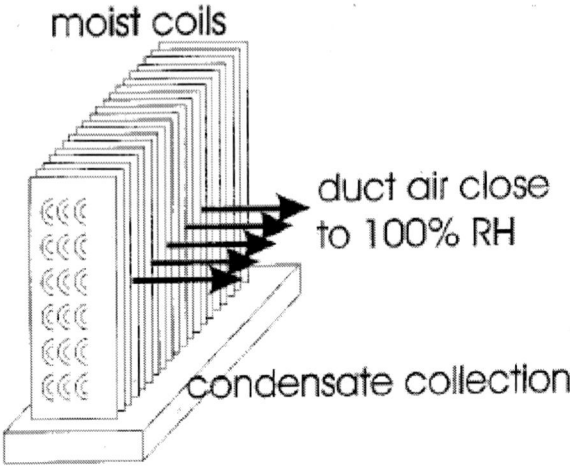

Figure 29. Damp Duct Symptoms (Chamra et al., 2000).

Summary of Principles of Desiccant Systems

Desiccant dehumidification systems remove moisture from the air by forcing the water vapor directly into a desiccant material. The moisture from the air is attracted to desiccants since an area of low vapor pressure is created at the surface of the desiccant. The pressure exerted by the water in the air is higher, so the water molecules move from the air to the desiccant and the air is dehumidified.

The functioning of desiccant material might be compared to the action of a sponge in collecting a liquid. When the sponge is dry, it soaks up the liquid effectively. Once it becomes saturated, the sponge is taken to a different spot, the liquid is expelled by squeezing the sponge, and the sponge is ready to absorb more liquid. In a desiccant system, if the desiccant material is cool and dry, its surface vapor pressure is low, and moisture is attracted and absorbed from the air, which has a higher vapor pressure. After the desiccant material becomes wet and hot, it is moved to another air stream and the water vapor is expelled by raising the temperature (this step is called "regeneration"). After regeneration, the desiccant material is ready to be brought back to absorb more water vapor. Unlike the conventional cooling coil, the water vapor does not condense, but rather remains a vapor throughout all processes.

Desiccants can be either solids or liquids. The difference between solid and liquid desiccants is their reaction to moisture. Some simply collect moisture like a sponge collects water. These desiccants are called adsorbents and are mostly solid materials. Silica gel is an example of a solid adsorbent. Other desiccants undergo a chemical or physiological change as they collect moisture. These are called absorbents and are usually liquids or solids, which become liquid as they absorb moisture. Lithium chloride collects water vapor by absorption. Sodium chloride, common table salt, is another example of an absorbent.

TYPES OF DESICCANT SYSTEMS

General Classifications

Most commercial desiccant dehumidification systems use as their working material either a solid adsorbent or a liquid absorbent. Briefly, absorption is a process in which the nature of the absorbent is changed, either physically, chemically, or both. The change may include formation of a hydrate or phase change. An adsorbent, on the other hand, does not change either physically or chemically during the sorption process.

A variety of factors dictate whether an adsorbent will be commercially useful. These include cost, long-term stability, moisture removal characteristics (rate, capacity, saturation conditions, suitable temperatures), regeneration requirements (rate of moisture surrender as a function of temperature and humidity), availability, and manufacturing considerations.

Solid Adsorbents

Silica gels and zeolites are used in commercial desiccant equipment. Other solid desiccant materials include activated aluminas and activated bauxites. The desiccant material choice for a particular application depends on factors such as the regeneration temperature, the level of dehumidification, and the operating temperature.

Solid desiccant materials are arranged in a variety of ways in desiccant dehumidification systems. A large desiccant surface area in contact with the air stream is desirable. A way to bring regeneration air to the desiccant material is necessary.

The most common configuration for commercial space conditioning is the desiccant wheel shown in Figure 30a. The desiccant wheel rotates continuously between the process and regeneration air streams. The wheel is constructed by placing a thin layer of desiccant material on a plastic or metal support structure. The support structure, or core, is formed so that the wheel consists of many small parallel channels coated with desiccant. Both "corrugated" and hexagonal (Figure 30b) channel shapes are currently in use. The channels are small enough to ensure laminar flow through the wheel. Some kind of sliding seal must be used on the face of the wheel to separate the two streams. Typical rotation speeds are between 6 and 20 revolutions per hour. Wheel diameters vary from one foot to over twelve feet. Air filters are an important component of solid desiccant systems. Dust or other contaminants can interfere with the adsorption of water vapor and quickly degrade the system performance. All commercial systems include filters and maintenance directions for keeping the filters functioning properly.

Liquid Absorbents

Some materials that function as liquid absorbents are ethylene glycol, sulfuric acid, and solutions of the halogen group such as lithium chloride, calcium chloride, and lithium bromide (ASHRAE Fundamentals Handbook, 2001). A generic configuration for a liquid desiccant system is illustrated in Figure 31. The process air is exposed to a concentrated desiccant solution in an absorber, usually by spraying the solution through the air stream. As the solution absorbs water vapor from the air stream, the solution concentration drops, and the weak solution is taken to a regenerator where heat is used to drive off the water (which is carried away by a regeneration air stream) and the concentrated solution is returned to the absorber.

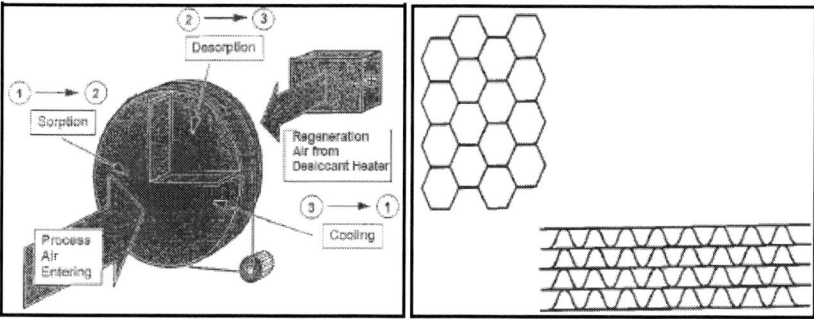

Figure 30. (a) Desiccant Wheel (Meckler et al., 1995) (b) Corrugated and Hexagonal Channel Shapes (Chamra et al., 2000).

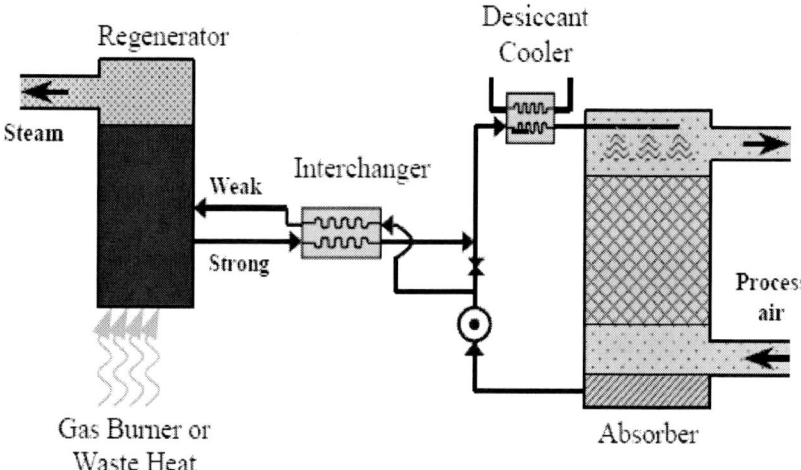

Figure 31. Liquid Desiccant System (Chamra et al., 2000).

Regeneration

For solid or liquid systems, regeneration energy can be drawn from a variety of sources. In an micro-CHP system, regeneration energy is drawn from the waste heat of a power-generation component. Due to the relatively low temperature requirements of regeneration (< 250 F or < 120 C), waste heat provided by combustion turbines, IC engines, and any of the fuel cell technologies is capable of supplying heat at regeneration temperatures. The thermal energy produced in many micro-CHP systems is sufficient to meet the input requirements for absorption refrigeration as well as desiccant regeneration.

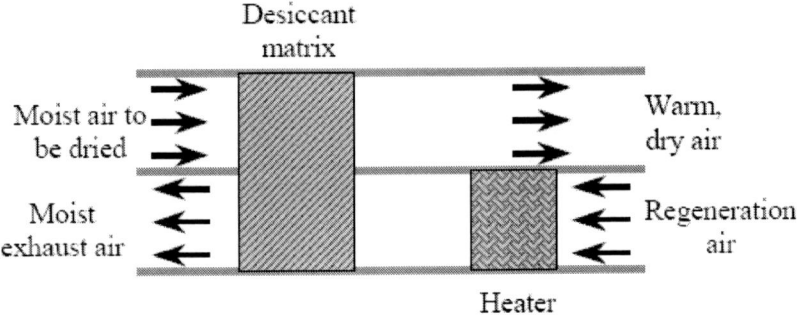

Figure 32. Solid Desiccant Dehumidification System (Chamra et al., 2000).

Solid Desiccant Systems

Figure 32 illustrates the components of a generic solid desiccant dehumidification system. At a minimum, the system will include separated process and regeneration air-streams for the desiccant device and some kind of heater to raise the temperature of the regeneration air.

The approximate path of the process air through a desiccant device is shown in Figure 33 for the same inlet and outlet conditions as were shown for the sub-cooling system (Figure 27). Note that, as indicated by the path from point 1 to point 2 in Figure 33, the desiccant process increases the dry bulb temperature of the process air. For solid desiccant materials, this increase is a result of the "heat of adsorption" which consists of the latent heat of vaporization of the adsorbed liquid plus an additional "heat of wetting." Heat of wetting is the energy released during dehumidification, in excess of the latent heat of vaporization. The path from point 1 to point 2 is close to a line of constant enthalpy. After the dehumidification process, the process air must undergo a sensible cooling process to reach the end point.

Cost Considerations

In many cases, the additional benefits provided by a desiccant system will lead to greater overall capital equipment cost. However, since the latent part of the cooling load is shifted from electrical energy to thermal energy, desiccant dehumidification systems can potentially have lower operating costs, particularly if waste heat can be utilized. In micro-CHP systems, waste heat can be used to regenerate the desiccant. This makes desiccant dehumidification appealing to micro-CHP applications, especially when an absorption chiller is incorporated into the system to provide the sensible cooling load.

Because the humidity and temperature can be controlled independently with a desiccant dehumidification and cooling system, the system performance is often more effective than that obtainable with conventional systems. Analysis of the sensible heat ratio (SHR) suggests the energy cost savings potential that a desiccant system may have. The SHR is the ratio of sensible cooling load to the total cooling load (sensible load plus latent load). A sensible heat ratio close to unity implies that very little moisture is removed from the air, while a sensible heat ratio close to zero indicates that most of the load is latent cooling. Air-conditioned environments often have SHR values well below unity, which results in greater energy consumption for a sub-cooling system than that of a desiccant dehumidification and cooling system that meets the same temperature/humidity requirements.

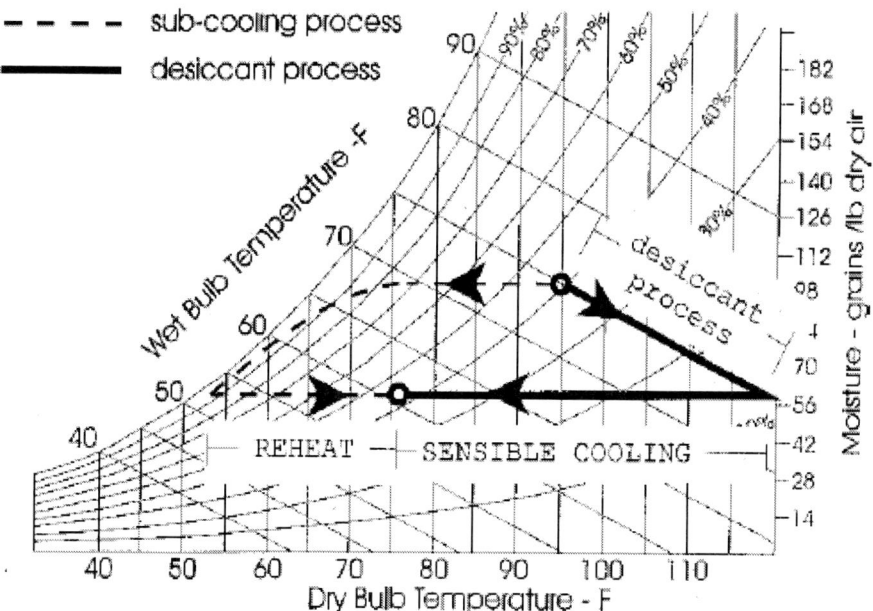

Figure 33. Dry Desiccant Dehumidification Process (Chamra et al., 2000).

Chapter 5

CONCLUSIONS

Micro-CHP is a developing paradigm in energy systems. The driving potential behind micro-CHP systems is the thermal efficiency that these systems can achieve and the significant portion of the energy market that lies within the "micro" regime. Projected system efficiencies of 80% are well above the overall thermal efficiencies produced by standard energy systems. micro-CHP systems have the potential to increase the overall thermal efficiency, reduce the total power requirement, and provide higher quality, more reliable power than conventional systems. Larger homes, higher energy costs, volatile fuel markets, electricity blackouts, power security, power quality, and lower emissions are additional characteristics that make micro-CHP attractive.

A variety of distributed power generation (DPG) technologies may be selected for an micro-CHP system. Table 11 illustrates how these technologies compare in efficiency, cost, technology status, emissions, noise, and load matching flexibility. The technologies are ranked from those having the most positive characteristics to those having the most negative characteristics. For example, fuel cells have the quietest operation and, therefore, the best characteristics in the noise category, and reciprocating engines generate significant noise and, therefore, the most negative characteristics in this category. The rankings in Table 11 are based on the technologies as a whole and may vary in some instances.

Table 11. Rankings for Distributed Power Generating Technologies

Rankings for DPG Technologies	
Category	Most Positive ⇒ Least Positive
Efficiency	Fuel Cells⇒ Reciprocating⇒ Stirling⇒ Microturbine⇒ Rankine
Technology Status	Reciprocating⇒ Microturbine⇒ Fuel Cells⇒ Stirling⇒ Rankine
Cost	Reciprocating⇒ Microturbine⇒ Fuel Cells⇒ Stirling⇒ Rankine
Emissions	Fuel Cells⇒ Stirling⇒ Microturbine⇒ Rankine⇒ Reciprocating
Noise	Fuel Cells⇒ Stirling⇒ Microturbine⇒ Rankine⇒ Reciprocating
Load Matching Flexibility	Reciprocating⇒ Fuel Cells⇒ Microturbine⇒ Stirling⇒Rankine

The thermally activated components discussed in this chapter are non-competing technologies. The selected use of either an absorption chiller or desiccant dehumidifier is based upon the needs of the application rather than performance. A system configuration requiring only latent cooling would incorporate an absorption chiller. If both cooling and dehumidification are required, both a desiccant dehumidifier and absorption chiller can be included in an micro-CHP system.

Despite the potential benefits of micro-CHP systems, numerous significant obstacles and challenges must be addressed if micro-CHP is to be successful in the United States. Research must be conducted to validate micro-CHP systems and components. Likewise, efforts must be made to reduce installed initial capital costs of micro-CHP systems. These issues can be resolved with time and the application of good engineering practices. The more difficult issues to overcome are characteristics of the market to which micro-CHP systems apply.

In order for micro-CHP systems to succeed, maintenance and service support issues must be addressed. Large scale CHP systems are typically installed in large factories, hospitals, or other facilities which employ a full-time maintenance staff. Retaining a full-time maintenance staff for a single micro-CHP system is neither practical nor economically feasible. Expecting the owner to maintain the system is also impractical because of technical complexity and safety issues. Still, providing professional, timely service and maintenance will be a crucial part of successful micro-CHP system operation. More than one potential technology has become a distant memory because of poor maintenance and technical support. This situation leads to a very important question: Who will provide maintenance and support? Should this responsibility fall on utility providers, individual micro-CHP system manufacturers, or independent maintenance contractors?

Another important issue to address is the inclusion and cooperation of local electrical utilities. Ideally, an micro-CHP system would provide the exact amount of electrical power needed instantaneously, continually throughout the day. While designing such a system is possible, unfortunately, the increase in system complexity and increased capital costs would be a detriment to early system installations. Initially, most micro-CHP systems will likely be designed as constant-power output or base-load systems. This implies that at some point the power requirement will not be met, or that the requirement will be exceeded. Realistically, both cases will occur within a 24-hour period. For example, in the United States, the base electrical load for the average home is approximately 2 kW while the peak electrical demand is slightly over 4 kW. (EIA) If a 3 kWe micro-CHP system were installed in this situation, part of the time more energy will be provided than could be used and for a portion of the time more energy will be required than could be provided.

One option is to size the system for base electrical load. This option requires that the system include an energy storage device or be connected to the electrical grid to provide the peak electrical load. Another option is to size the system for peak electrical load. This option requires that the excess energy be stored, transferred to the electrical grid, or simply rejected as a loss. In all instances, either energy storage or grid connectivity is required. Energy storage is expensive and increases the system size and initial costs. In many instances, a more economically attractive option is for the micro-CHP system to be grid connected.

Grid connectivity will require the inclusion and cooperation of local electrical utilities to establish an infrastructure and pricing system to accommodate micro-CHP installations. Grid connectivity also requires the use of two-way metering. The Institute of Electrical and Electronics Engineers-United States of America (IEEE-USA) has developed a series of standards for interconnection of distributed energy resources with the electricity grid. The IEEE 1547 2003 Standard for Interconnecting Distributed Resources With Electrical Systems is the first in the 1547 series of planned interconnection standards. The standard establishes technical requirements for electrical power systems (EPS or electrical grids) interconnecting with distributed generators such as fuel cells, photovoltaics, microturbines, reciprocating engines, wind generators, large turbines, and other local generators. Additional IEEE interconnection standards activities are now designated under the 1547 series of standards.

Before a full-market drive for installation of micro-CHP systems is attempted, an in-depth case study of an micro-CHP system should be performed. Also, a full economic assessment including all capital costs and the potential effects of the volatile fuel market should be conducted. The selection and price of fuel to be used in an micro-CHP system is very important. A small percent change in the market demand creates a large change in the cost of many fuels.

REFERENCES

ASHRAE, 2001, *2001 ASHRAE Handbook – Fundamentals*, the American Society of Heating, Refrigerating and Air-Conditioning Engineers, Inc., Atlanta, Georgia.

California Distributed Energy Resources Guide. *Available at http://www.energy.ca.gov/distgen/*

Caterpillar. Available at http://www.cat.com

Capstone. Available at www.capstone.com

Chamra, Louay, Parsons, Jim A., James, Carl, Hodge, B.K., and Steele, W. Glenn, Desiccant Dehumidification Curriculum Module for Engineering/Technology HVAC Courses, Mississippi State University, 2000.

DOE Energy Information Administration. *Available at http://www.eia.doe.gov.*

DOE Industrial Technologies Program. *Available at http://www.oit.doe.gov.*

Georgia State University Physics Website. *Available at http://hyperphysics.phy-astr.gsu.edu/hbase/thermo/diesel.html.*

Harold, Keith E., Klein, Sanford A., and Radermacher, Reinhard. Absorption Chillers and Heat Pumps, New York: CRC Press, Inc., 1996.

Hewitt, G. F., Shires, G. L. and Bott, T. R. "Process Heat Transfer," CRC Press., 1994.

Keveny, Matt, 2000, "Two Cylinder Stirling Engine." *Available at http://www.keveny.com/Vstirling.html*

Knight, I., and Ugursal, I., *Residential cogeneration systems: a review of the current technologies.* A report of Subtask A of FC+COGEN-SIM The Simulation of Building-Integrated Fuel Cell and Other Cogeneration Systems" Annex 42 of the International Energy Agency Energy Conservation in Buildings and Community Systems Programme. 93 pages. First published: April 2005

Laraminie, James, and Dicks, Andrew, Fuel Cell Systems Explained, 2nd edition, West Sussex: Wiley, 2003.

Meckler, M., R. Heimann, J. Fischer, and K. McGahey. Desiccant Technology Transfer Workshop Manual. American Gas Cooling Center, 1995, Arlington, Virginia.

Micro Cogeneration system for residential and small commercial applications. *Available at www.cogenmicro.com*

Plug Power. *Available at www.plugpower.com*

RS Means Mechanical Cost Data 24th Annual Edition, Kingston, Maryland: RS Means Company Inc. 2001

Shah, Ramesh K., "Recuperators, Regenerators and Compact Heat Exchangers," CRC Handbook of Energy Efficiency, New York: CRC Press, Inc., 1997.

"The micro-CHP Technologies Roadmap: Meeting 21st Century Residential Energy Needs" December 2003 United States Department of Energy Office of Energy Efficiency and Renewable Energy

Solo Stirling Engine. *Available at www.stirling-engine.de/engl/*

The U.S. Department of Defense (DoD) Fuel Cell Test and Evaluation Center. *Available at http://www.fctec.com/fctec*

Van den Oosterkamp, P.F., Goorsen, A.J., and Blomen., L.J., *Journal of Power Sources*, 1993 41, p.239 - 252.

ADDITIONAL REFERENCES

Alanne, Kari, and Saari, Arto. Sustainable small-scale CHP technologies for buildings: the basis for multi-perspective decision making. *Renewable & Sustainable Energy Reviews*, Helsinki University of Technology, Hut, Finland.

Basso, T. S. and DeBlasio, R., IEEE 1547 series of standards: interconnection issues. *IEEE Transactions on Power Electronics,* 2004 Vol 19, 5.

Boyce, Meherwan P., Handbook for Cogeneration and Combined Cycle Power Plants, New York: ASME Press, 2002.

Caton, Jerald A., and Turner, W. Dan, "Cogeneration," CRC Handbook of Energy Efficiency, New York: CRC Press, Inc., 1997.

Cengel, Yunus A., Heat Transfer: A Practical Apprach, 2nd edition, New York: McGraw Hill, 2003.

Commercial micro-CHP Using Fuel Cells and Microturbines. *Emerging Technologies and Practices: 2004 American Council for an Energy- Efficient Economy.*

References

Felder, Richard M., and Rausseau, Ronald W., ElementaryPrinciples of Chemical Processes, 2nd edition, New York: John Wiley and Sons, 1986.

Flin, D. Domestic CHP in Europe. *Cogeneration and On-Site Power Production*, Jan. – Feb. 2005, pp. 43 – 49.

Goswami, D.Y., Kreith, F., and Kreider, J., Principles of Solar Engineering, 2nd edition, Taylor and Francis Pub., 2000.

Hardy, J.D, Cooling, Heating, and Power for Building Instructional Module, Mississippi State University, 2004.

Harrison, J. Micro CHP in rural areas. *EarthScan James & James*. 1 January, 2003a.

Harrison, J. Micro CHP in Europe. *EA Technology*. Presentation at the 2003 National Micro CHP Technology Pathways Workshop. 2003b.

Harrison, J. Towards a strategy for micro CHP in the USA domestic markets. *EA Technology,* June 2003c.

Moran, Michael J., and Shapiro, Howard N., Fundamentals of Engineering Thermodynamics, 4th edition, New York: John Wiley and Sons, 2000.

Micro Gas Turbines and Heat-Driven Cooling. *Australian National Training Authority*, 2003.

Peltchers, Neil, Combined Heat and Power Handbook: Technologies and Applications, The Fairmont Press, Inc., Georgia, 2003.

Pehnt, P., Praetorius, B., Foscher, C., Schnieder, L., Cames, M., and VoB, J.P. Micro CHP – a sustainable innovation. *Transformation and Innovation in Power Systems*. Berlin/Heidelberg, 2004.

Stull, D.R., et al., JANAF Thermochemical tables, Michigan: Dow Chemical Company, 1965.

Senft, James R., Ringbom Stirling Engines, New York: Oxford University Press, 1993.

Urieli, Isreal, and Berchowitz, David M., Stirling Cycle Engine Analysis, Great Britain: Adam Hilger Ltd.,1984.

INDEX

A

absorbents, 50, 61, 62
absorption, 1, 4, 11, 15, 16, 38, 46, 47, 49, 50, 51, 52, 53, 54, 55, 56, 61, 64, 65, 68
AC, 23, 27
acid, 23, 24, 25, 27, 34, 35, 37, 62
activation, 26
adsorption, 62, 64
AFC, 27
aid, 30
air, vii, 1, 4, 10, 16, 17, 18, 20, 27, 30, 31, 32, 42, 46, 49, 52, 53, 57, 58, 59, 60, 61, 62, 64, 65
air quality, 10
air travel, 59
aircraft, 14, 18
alkali, 35
alkaline, 27
alpha, 18
alternative, 5
ambient air, 49, 52
ammonia, 52, 53, 55
anode, 24, 25, 27, 29, 30, 31, 32, 33, 34, 35, 36
antimony, 26
application, 7, 9, 11, 18, 37, 61, 68
arsenic, 26
assessment, 69
atmosphere, 11, 14, 20
atoms, 29
attention, 5
automobiles, 37
automotive, 38
availability, 13, 61

B

barriers, 29
batteries, 13, 23
behavior, 17
benefits, 64, 68
biofuels, 9, 14, 17
black, 34
blackouts, 5, 67
boilers, 47
boiling, 50, 51, 52
boils, 52
BOP, 13
Britain, 73
Bromide, 54
buildings, vii, 1, 7, 14, 38, 52, 57, 72
burning, 18, 20
buses, 15

Index

C

calcium, 62
California, 71
candidates, 27
capacity, 14, 19, 54, 61
capital, 9, 12, 16, 17, 20, 23, 56, 64, 68, 69
capital cost, 9, 12, 16, 17, 20, 56, 68, 69
carbon, 23, 31, 33, 34, 35, 37
carbon dioxide, 23, 31, 33, 35, 37
carbon monoxide, 31, 33
carrier, 37
case study, 69
catalyst, 9, 23, 27, 32, 34, 36, 39
catalytic, 24
cathode, 24, 25, 26, 27, 29, 30, 31, 32, 33, 34, 35, 37
cell, 23, 24, 25, 26, 27, 29, 30, 32, 33, 35, 37, 38, 39, 40, 41, 63
ceramic, 30, 32, 35
CH4, 28
channels, 62
chemical, 23, 25, 34, 41, 42, 61
chemical energy, 41
chemistry, 37
chloride, 61, 62
CHP, v, vii, 1, 2, 4, 5, 7, 9, 11, 13, 14, 15, 17, 18, 20, 21, 22, 23, 27, 29, 30, 31, 32, 33, 35, 38, 42, 46, 47, 49, 55, 58, 63, 65, 67, 68, 69, 72, 73
classical, 44
classification, 42
classified, 42, 43, 46, 49, 54
closed-loop, 22
clusters, 14
CO2, 28, 36
coal, 3
coefficient of performance (COP), 52
coil, 46, 58, 59, 61
combustion, 9, 11, 12, 13, 16, 17, 20, 33, 39, 41, 42, 46, 63
commercial, vii, 1, 2, 7, 14, 15, 21, 32, 37, 38, 52, 55, 57, 61, 62, 72
communities, vii
complexity, 68, 69
components, vii, 1, 4, 7, 9, 16, 23, 26, 27, 29, 31, 41, 42, 46, 49, 51, 64, 68
compression, 9, 17, 49, 50, 51, 52, 56
concentration, 26, 53, 62
condensation, 12, 43, 49, 51, 52
conditioning, 1, 32, 42, 52, 53, 58, 62
conductive, 35
conductivity, 29
configuration, 15, 55, 62, 68
Congress, iv
connectivity, 69
constraints, 41
construction, 21, 27, 31, 32, 36
consumption, 1, 2, 5, 65
contaminants, 62
contractors, 68
contracts, 18
control, 29, 32, 49, 59
controlled, 20, 29, 37, 65
convection, 46
conversion, 2, 9, 20, 21, 23, 26, 30, 41
cooling, vii, 1, 4, 7, 9, 11, 12, 15, 16, 20, 36, 38, 41, 49, 57, 58, 59, 61, 64, 65, 68
cooling process, 58, 64
COP, 55
corrosion, 21, 37, 57
corrosive, 12
cost saving, 65
costs, 5, 11, 12, 13, 15, 16, 17, 18, 20, 21, 23, 27, 29, 30, 32, 39, 41, 56, 65, 67, 68, 69
coupling, 55
CRC, 45, 71, 72
customers, 3

D

dating, 5
decentralized, vii, 1
decision making, 72
degradation, 57
delivery, 3
Delphi, 39
demand, 13, 69
density, 40, 41
Department of Defense, 72

Index

Department of Energy, 72
deregulation, vii
desire, 38
Dicks, 72
diesel, 9, 10, 13, 37, 39, 71
diesel fuel, 9
diffusion, 29
Dil, 28
distributed generation, 10, 38, 46
distribution, 3, 5
district heating, 46
DOE, 5, 39, 71
domestic markets, 73
dry, 58, 60, 64
durability, 21

E

EA, 73
eating, 9
economic, 29, 69
EIA, 69
electric circuit, 25
electric energy, 49
electric power, 3, 4, 7, 11, 20, 21, 41
electric utilities, vii
electrical, vii, 1, 3, 5, 7, 9, 13, 17, 19, 20, 21, 23, 26, 29, 30, 32, 33, 34, 41, 65, 68, 69
electrical power, vii, 1, 3, 5, 9, 23, 41, 69
electricity, vii, 1, 3, 4, 5, 24, 27, 29, 32, 35, 39, 41, 67, 69
electrocatalyst, 34
electrochemical, 23, 34
electrodes, 29
electrolyte, 24, 25, 27, 29, 30, 31, 32, 34, 35, 37
electronic, iv
electrons, 24, 25, 29, 31, 35, 36
electrostatic, iv
endurance, 35
energy, 1, 2, 3, 4, 5, 7, 9, 10, 11, 12, 13, 14, 15, 17, 18, 19, 21, 23, 25, 30, 37, 39, 40, 41, 42, 47, 49, 51, 54, 59, 63, 64, 65, 67, 69, 71
energy consumption, 1, 2, 65

energy density, 40
Energy Efficiency and Renewable Energy, 72
Energy Information Administration, 71
engineering, vii, 13, 68
engines, 7, 9, 10, 11, 12, 13, 14, 16, 17, 18, 20, 21, 22, 41, 46, 47, 63, 67, 69
England, 18
entropy, 25, 26
environment, 32, 50, 52
environmental, vii, 5
environmental issues, 5
equipment, 7, 9, 10, 13, 14, 23, 39, 40, 59, 61, 64
ER, 11
ethylene, 62
ethylene glycol, 62
Europe, 18, 19, 73
European Commission, 32
evaporation, 43
exhaust heat, 12
exothermic, 24
expenditures, 1
expert, iv

F

fabrication, 32, 35
fees, 13
feet, 62
filters, 62
Finland, 72
fire, 46
first generation, 34
flexibility, 15, 19, 23, 26, 67
float, 47
flood, 29
flow, 35, 43, 44, 46, 47, 62
flue gas, 46
fluid, 20, 21, 22, 42, 43, 46, 47, 55
fluidized bed, 47
fossil, 15
free energy, 25, 32
freezing, 52

fuel, 2, 3, 5, 7, 9, 12, 13, 15, 17, 18, 19, 20, 21, 23, 24, 25, 26, 27, 29, 30, 31, 32, 34, 35, 36, 37, 38, 39, 40, 41, 63, 67, 69
fuel cell, 23, 24, 25, 26, 27, 29, 30, 31, 32, 34, 35, 36, 37, 38, 39, 40, 41, 63, 67, 69
fuel type, 13

G

gas, 3, 9, 10, 12, 13, 14, 15, 17, 19, 24, 26, 29, 30, 32, 33, 35, 36, 37, 38, 39, 46, 47, 53, 55
gas diffusion, 29
gas turbine, 17, 39
gases, 9, 11, 15, 17, 20, 23, 29, 41, 42, 46, 47, 55
gasoline, 9
gel, 61
gels, 61
generation, vii, 1, 3, 5, 7, 9, 10, 11, 14, 17, 18, 20, 23, 29, 34, 38, 39, 41, 42, 46, 54, 63, 67
generators, 16, 17, 39, 42, 69
Georgia, 71, 73
Gibbs free energy, 25, 32
glass, 58
glycol, 62
government, 38, 39
grains, 57, 58
Great Britain, 73
Grid connectivity, 69
grids, 5, 69
growth, 57, 60

H

H_2, 28
halogen, 62
harmful, 53
harvesting, 2
hazards, 35
health, 57, 60
heat, vii, 1, 3, 4, 5, 7, 9, 11, 12, 15, 16, 17, 20, 21, 22, 23, 27, 30, 32, 33, 35, 36, 38, 40, 41, 42, 43, 44, 46, 47, 49, 51, 52, 53, 54, 55, 56, 58, 62, 63, 64, 65

heat exchangers, vii, 1, 17, 41, 42, 43, 44, 45, 46, 47, 72
heat loss, 20
heat pumps, 49
heat transfer, 42, 44, 46, 47, 51
heating, vii, 1, 4, 7, 9, 11, 13, 15, 16, 31, 32, 34, 36, 41, 42, 46, 47, 49
high power density, 41
high pressure, 12, 49
high temperature, 11, 27, 32, 37, 38, 49, 52, 54
higher quality, 1, 67
homes, 5, 58, 67
hospitals, 38, 57, 68
hot water, vii, 1, 9, 11, 12, 20, 31, 36, 37, 38, 41, 54, 55
hotels, 15, 57
household, 1, 5
hub, 3
humidity, 29, 49, 57, 58, 59, 61, 65
hybrid, 15
hydrate, 61
hydro, 3, 26
hydrocarbons, 26
hydrogen, 13, 23, 24, 26, 27, 29, 30, 31, 33, 35, 37
hydrogen gas, 24

I

ice, 57, 58
inclusion, 68, 69
incomes, 5
indication, 25
industrial, 1, 5, 35, 37, 38, 47
industrial application, 38, 47
industrialized countries, 5
industry, vii, 5
inertia, 30
infrastructure, 11, 69
injury, iv
innovation, 73
inorganic, 34
inspection, 13, 39
internal combustion, 11, 12, 17

International Energy Agency, 71
ions, 24, 25, 27, 29, 31, 35, 36, 37

J

January, 39, 73

L

lamina, 62
laminar, 62
landfill gas, 10, 15, 37
lanthanum, 30
large-scale, 7, 11, 20, 35
lead, 37, 64
light industrial, 15, 38
liquids, 41, 42, 50, 61
lithium, 35, 52, 54, 55, 61, 62
location, 10
long-term, 39, 61
losses, 2, 3, 5, 20, 26, 41
low temperatures, 27
low-temperature, 7, 42, 46, 52
lubricating oil, 16
lubrication, 16, 20

M

machines, 52
magnetic, iv
maintenance, 11, 13, 14, 15, 16, 17, 21, 23, 29, 30, 39, 56, 62, 68
Maintenance, 13, 39
management, 29, 32
manufacturer, 20, 27
manufacturing, 61
market, vii, 1, 2, 5, 16, 29, 30, 32, 35, 38, 67, 68, 69
market trends, vii
market value, 30
marketing, 11
markets, 5, 37, 67, 73
Maryland, 72
matrix, 35

mechanical, iv, 7, 11, 23, 41, 56
media, 52
melt, 35
memory, 68
methane, 26, 30
methanol, 27
MFC, 27
microbial, 60
missions, 53
Mississippi, 71, 73
mixing, 43, 51
mobility, 35
models, 10, 19
moisture, 57, 58, 60, 61, 65
moisture content, 58
mole, 36
molecules, 31, 60
motels, 15
motion, 17
motivation, 3
motors, 5

N

natural gas, 3, 9, 13, 14, 29, 30, 32, 33, 35, 36, 38, 55
NC, 32
neglect, 13
Netherlands, 30, 32
New York, iii, iv, 71, 72, 73
nickel, 30, 32, 36
noise, 10, 15, 17, 21, 41, 67
normal, 46, 47
nuclear, 3

O

oil, 11, 13, 16, 37, 47
organic, 47
Ottawa, 32
oxide, 23, 27, 30, 31, 32, 38
oxygen, 23, 24, 26, 27, 29, 30, 31, 35

P

PAFC, 23, 25, 27, 28, 34, 35, 36, 37, 38, 40, 56
particles, 34, 47
partnership, 39
payback period, 30, 56
PEMFC, 23, 27, 28, 29, 30, 34, 35, 36, 37, 38, 39, 40, 56
performance, 21, 23, 25, 33, 35, 37, 46, 52, 62, 65, 68
periodic, 13, 39
Periodic Table, 26
petroleum, 3
PG, 67
phosphoric acid fuel cell, 25
phosphorous, 26
photovoltaics, 69
physiological, 61
plants, 3, 5, 11, 21, 42
plastic, 62
platinum, 27, 34
poisoning, 29, 39
poisons, 26
polarization, 26
pollution, vii, 3, 41
polymer, 27, 29
poor, 68
pores, 29
porous, 30
potassium, 35
power, vii, 1, 3, 4, 5, 7, 9, 11, 12, 13, 14, 15, 16, 17, 18, 20, 21, 22, 23, 26, 27, 29, 32, 37, 38, 39, 41, 42, 43, 55, 63, 67, 69
power generation, vii, 3, 5, 7, 9, 11, 14, 17, 18, 20, 23, 29, 38, 41, 42, 67
power plant, 3, 5, 20, 21, 22, 43
power plants, 3, 5, 20, 21, 22
power stations, 17
premium, 11
preparation, iv, 16
pressure, 11, 12, 13, 21, 36, 38, 49, 50, 51, 52, 54, 60
producers, vii
production, 2, 11, 18, 21, 35, 39, 42
production costs, 18, 21, 39
program, 39
promote, 44
propane, 9, 14, 37
property, iv
protection, 57
proton exchange membrane, 23, 27
prototype, 55
PTFE, 27
pumps, 9, 49

R

R&D, 40
radiation, 11, 46
range, 9, 13, 14, 16, 17, 20, 30, 32, 38, 42, 52, 54
RC, 72
reactants, 25
recovery, 9, 12, 15, 16, 17, 22, 41, 42, 46, 47
reduction, 50
refrigerant, 49, 50, 51, 52, 53, 54
refrigeration, 18, 49, 52, 64
regenerate, 65
regeneration, 58, 60, 61, 62, 63, 64
regulations, vii
reliability, 9, 15, 16, 21, 30, 39, 40, 56
renewable energy, 3
research, 5, 17, 32
research and development, 17
researchers, 39
reservoir, 17
residential, vii, 1, 2, 4, 5, 7, 18, 29, 32, 37, 38, 41, 52, 55, 57, 72
residential buildings, 57
resources, 11, 69
response time, 37
restaurants, 15
restructuring, vii
retail, 15, 57
rice, 5
rural, 73
rural areas, 73

S

safety, 68
sales, 16
salt, 61
salts, 35
saturation, 61
savings, 65
seals, 20
security, 5, 67
seeds, 57
selecting, 26
selenium, 26
Self, 40
series, 43, 69, 72
services, iv
shape, 43
Shell, 45, 47
Siemens, 32
SME, 72
sodium, 35
SOFC, 23, 27, 28, 30, 31, 32, 33, 35, 37, 38, 39, 40, 56
software, 16
solar, 17, 18
solar energy, 17
solid state, 30
solutions, 62
sorption, 61
sorption process, 61
stability, 34, 61
stages, 54, 55
stainless steel, 36
standards, 10, 32, 69, 72
steel, 36
storage, 15, 33, 69
streams, 23, 43, 51, 62, 64
substances, 25, 50
substation, 3
sulfur, 26
sulfuric acid, 62
suppliers, 39
supply, 20, 26, 37, 39, 41, 55
surface area, 61
Synergy, 40

systems, vii, 1, 5, 7, 9, 11, 12, 15, 16, 20, 29, 32, 39, 41, 42, 46, 47, 49, 52, 54, 55, 56, 58, 60, 61, 62, 63, 65, 67, 68, 69, 71

T

tariffs, 30
technicians, 11
technological, vii
technology, vii, 1, 5, 9, 11, 12, 13, 15, 16, 18, 20, 23, 29, 32, 34, 41, 67, 68
Teflon, 27
tellurium, 26
temperature, 7, 11, 12, 15, 17, 25, 26, 27, 29, 30, 32, 34, 35, 36, 37, 38, 40, 41, 42, 43, 46, 47, 49, 51, 52, 54, 55, 58, 59, 60, 61, 63, 64, 65
theoretical, 25, 29, 30
thermal, 1, 4, 5, 7, 9, 11, 15, 18, 21, 26, 30, 32, 34, 41, 42, 47, 49, 51, 53, 55, 63, 65, 67
thermal efficiency, 1, 4, 7, 26, 41, 55, 67
thermal energy, 5, 9, 21, 42, 47, 49, 51, 63, 65
thermal load, 21
thermodynamic, 21
thermodynamic cycle, 21
thinking, vii
thresholds, 12
time, 1, 2, 3, 18, 30, 33, 37, 68, 69
toxicity, 52
training, 16
trans, 39
transfer, 36, 42, 43, 44, 46, 47, 51
transition, 5
transmission, 3, 5
transport, 27
transportation, 2, 15, 39
travel, 29, 35
trial, 23
trucks, 14
tubular, 32
two-way, 69

U

uncertainty, 5
unit cost, 11, 56
United States, 5, 68, 69, 72
universities, 38

V

values, 13, 65
vapor, 9, 13, 33, 49, 50, 51, 52, 53, 54, 55, 56, 57, 60, 61, 62
vehicles, 15, 18
velocity, 47
ventilation, 32, 53
vibration, 21
Virginia, 72
volatility, 34

W

warm air, vii, 1, 46
waste, vii, 1, 3, 4, 15, 17, 20, 27, 30, 32, 38, 40, 41, 42, 46, 47, 55, 56, 58, 63, 65
water, vii, 1, 4, 9, 11, 12, 13, 16, 20, 22, 23, 24, 29, 31, 32, 33, 35, 36, 37, 38, 41, 47, 52, 53, 54, 55, 56, 57, 58, 60, 61, 62
water absorption, 53
water heater, 47
water vapor, 13, 33, 53, 57, 60, 61, 62
wells, 2
Westinghouse, 32
wet, 58, 59, 60
wetting, 64
wind, 69

Y

yield, 55

Z

zeolites, 61
zirconia, 30